Power Density

Power Density

A Key to Understanding Energy Sources and Uses

Vaclav Smil

The MIT Press
Cambridge, Massachusetts
London, England

MIT Press books may be purchased at special quantity discounts for business or sales promotional use. For information, please email special_sales@mitpress.mit.edu.

This book was set in ITC Stone Sans Std and ITC Stone Serif Std by Toppan Best-set Premedia Limited. Printed and bound in the United States of America.

Library of Congress Cataloging-in-Publication Data

Smil, Vaclav.
Power density : a key to understanding energy sources and uses / Vaclav Smil.
  pages   cm
Includes bibliographical references and index.
ISBN 978-0-262-02914-8 (hardcover : alk. paper)
1. Energy level densities. 2. Energy auditing. 3. Renewable energy sources. 4. Power resources. I. Title.
TJ163.245.S65   2015
333.79–dc23
2014045995

10 9 8 7 6 5 4 3 2 1

Contents

Preface

Modern societies have many concerns about their energy supply. Above all, it should be affordable, reliable, and convenient: affordable in order to drive economic development and improvements in quality of life, reliable so as to be available on demand in its various forms, most of all as uninterruptible electricity, and convenient to give consumers virtually effortless access to preferred household, industrial, and transport energies. During the closing decades of the twentieth century two other concerns became prominent: energy supplies should also be environmentally benign and, preferably, renewable. Since the 1970s—that is, once energy matters began to receive, belatedly, an unprecedented amount of professional and public attention—all of these concerns have been addressed and analyzed in hundreds of books and thousands of papers, to say nothing of the instant expertise proffered by the mass media.

Any modestly informed person knows that energy prices matter (we need only think of OPEC's crude-oil price fixing), as do a reliable electricity supply (to avoid blackouts) and the finiteness of fossil fuel resources (underscored by claims about an imminent peak of oil extraction), and that the environmental impacts of energy use are not only local but worldwide (global warming). But does power density matter? Power is simply energy flow per unit of time (in scientific units, joules per second, which equals watts, or $J/s = W$), spatial density is the quotient of a variable and area, and hence power density is W/m^2, that is, joules per second per square meter. Why should we care about how much power is produced, or used, per unit of the Earth's surface? Some say explicitly that we should not bother at all.

Amory Lovins spent a lifetime making exceedingly optimistic forecasts about the speed with which renewable energy conversions (and other energy-related innovations) would be adopted by modern energy systems (for these claims, see Lovins 1977 and 2011a; for their critique, see Smil 2010a). In 2011 he dismissed any need to consider power densities, concluding that "land footprint seems an odd criterion for choosing energy systems: the amounts of land at issue are not large, because global renewable energy flows are so vast that only a tiny fraction of them need be captured" (Lovins 2011b, 40). In contrast, nearly two generations ago Wolfgang Häfele and Wolfgang Sassin, two leaders of energy research at the International Institute for Systems Analysis, wrote that "the density of energy operations is one of the most crucial parameters that predetermine the structure of the energy systems" (Häfele and Sassin 1977, 18).

Even so, most of the modern energy literature has simply ignored the subject. Energy economists have been worrying about prices, their elasticities, oligopolies, taxation, and links between energy use and economic growth. As interest in energy matters began to rise, Malcolm Slesser offered an explanation of this omission:

Land has always had a firm place in classical economics, where attention focused upon the agrarian economy, and land as a factor of production. Ricardo and Malthus founded their ideas around the land factor, in marked contrast to more recent economic thought in which land virtually dropped out of the scene, and production was viewed essentially as a synergy of labour and capital. (Slesser 1978, 86)

Similarly, modern engineering analyses of energy systems look at fuel qualities, mass throughputs, the rated power of energy converters, and annual and peak production. Of course, the design of specific energy extraction or conversion facilities must, inevitably, consider land requirements and land qualities in requisite detail, but space is not a common analytical denominator used to assess their performance. Still relatively uncommon interdisciplinary inquiries into the nature and linkages of modern energy systems look at per capita energy use, energy's role in economic performance, and the impacts of energy conversions on environmental quality, but only very few recent publications have looked at power densities.

I chose the measure as a key analytical variable in *General Energetics* (Smil 1991) and in its second, expanded edition, titled *Energy in Nature and Society* (Smil 2008). David McKay's *Sustainable Energy—Without the Hot Air* (2008) also uses the rate as an essential indicator that offers revealing insights into unfolding energy transitions. And Juan Moreno Cruz and M. Scott Taylor offer a rare example of economists who have explicitly incorporated energy density and power density in their new analysis, concluding that "perhaps the most important attribute of an energy source is its density: its ability to deliver substantial power relative to its weight or physical dimensions" (Cruz and Taylor 2012, 2) and that "the very density of the energy resource we seek, fuels our efforts to obtain more. Therefore differences in power density across energy resources create large differences in energy supply" (Cruz and Taylor 2012, 48).

In this book I demonstrate that power density is a key determinant of the nature and dynamics of energy systems. Its careful quantifications, their critical appraisals, and their revealing comparisons bring a deeper understanding of the ways with which we harness, convert, and use energies. A careful assessment of power densities is particularly revealing when contrasting our dominant fossil fuel–based energy system with renewable energy conversions. But before I start laying the foundation for my systematic power density assessments (by introducing the principal variables and reviewing different power density concepts), I tell a story of early modern charcoal-based iron smelting and of its replacement by coke-based production in blast furnaces. This historical appraisal offers a compelling (yet curiously overlooked) example of why power densities matter, why energy's rate of flow per unit of surface area is a key determinant of the structure of energy systems, and how it constrains their development.

Acknowledgments

All of my books written between 2000 and 2009 had, besides many photographs, numerous original illustrations on whose creation I worked with Doug Fast, a photographer and graphic artist at the University of Manitoba. Doug left the university in 2009, and as a result, all of my post-2010 books had either only rudimentary graphs or no original illustrations (just some photographs evoking the contents of individual chapters), and *Making the Modern World*, the last book to come out, in 2013, had no images at all.

Fortunately, this has changed, thanks to the assistance of Ian Saunders, who served as the project's creative director, and a team that included Anu Horsman, Jen Krajicek, Luke Shuman, Carl de Torres, and Wendy Woska. Additionally, Noli Novak created stippled portraits from low-resolution photographs of John Henry Poynting and Nikolai Alekseevich Umov. My thanks to all: as a lifelong graphics aficionado I greatly appreciate the difference simple, elegant illustrations can make.

1 How Power Density Matters

This brief chapter starts with a 1548 King's Commission, proceeds to a great English invention of the eighteenth century (not James Watt's steam engine!), and ends with the consequences of trying to supply a rising pig iron output in nineteenth-century blast furnaces by metallurgical charcoal.

In November 1548 the King's Commission, given to 20 men in Sussex, was to examine "the hurts done by iron mills and furnaces made for the same" on the Weald, the region of southeastern England between South and North Downs that was the center of English iron-making during the sixteenth century. The commission's most pressing questions to witnesses were these (Straker 1969, 115):

5. If the said iron mills and furnaces be suffered to continue, then whether thereby there shall be great lack and scarcity of timber and wood in parts near the mills?
6. What number of towns are like to decay if the iron mills and furnaces be suffered to continue?

The cause of these hardships was obvious:

The iron mills and furnaces do spend yearly ... above 500 loads of coals, allowing to every load of coals at the least three loads of wood; that is every iron mill spendeth at the least yearly 1,500 loads of great wood made into coals.

The petitioners went to great length to enumerate how wood scarcity would make it impossible to build houses, water mills or windmills, bridges, sluices, ships, boats, wheels, arrows, hogsheads, barrels, buckets, saddletrees, bowls, dishes and to supply timber "for the King's Majesty's towns and pieces on the other side of the sea" (Straker 1969, 118; Boulogne was an English possession between September 1544 and March 1550). As long as the deforestation caused by the rising demand for charcoal was limited to a few counties, local smelting would simply decline or cease, and the production would move elsewhere. Data assembled by King (2005) show that the production of pig iron from Weald's charcoal furnaces (about 4,000 tonnes [t] at the time of the 1548 Sussex petition) peaked by 1590 (at 14,040 t), and that by 1660 the region was making less iron than in 1550, while the metal's output kept on increasing in the rest of England.

English Iron Smelting

By 1620 England was producing more than 26,000 t of pig iron annually, but then production began to fall, and it was only about 18,000 t by the end of the seventeenth century. We can reconstruct fairly reliably what this meant in terms of energy demand and environmental impact. In the early decades of the eighteenth century, charcoal-fueled blast furnace campaigns usually extended from October to May, and during those eight months a typical furnace produced 300–340 t of pig iron (Hyde 1977). The efficiencies of wood conversion to charcoal and the charcoal requirements for pig iron smelting and for subsequent conversion of the metal to bar iron varied considerably (and about a third of the metal was wasted in the conversion), but the best available data (Hammersley 1973) indicate that at the beginning of the eighteenth century, at least 32 t of wood were needed to produce a tonne of bar iron.

This means that a typical furnace and an associated bar forge (sometimes located closer to the market) would have required at least 9,600–10,900 t of wood—and in the early seventeenth century, with less efficient conversions, the total could easily have been twice as much. The preferred wood supply came from coppiced plantings of beech or oak cut in 20-year rotation, which would yield an annual increment of about 5 m^3/ha; with an average wood density of 0.75 g/cm^3, that would be 3.75 t/ha, and operating a furnace and a forge would have required harvesting about 2,700 ha/year. In 1700 British furnaces produced about 12,000 t of bar iron, and hence they consumed on the order of 400,000 t of charcoaling wood. With an average productivity of 4 t/ha, this amount of wood would have required at least 100,000 ha of coppiced growth, a square with sides of nearly 32 km.

Obviously, the availability of suitable wood was a limiting factor in the further expansion of the English iron industry. Hammersley (1973) estimated that the maximum countrywide harvest would have been on the order of 1 Mt/year. Not surprisingly, during the eighteenth century England became highly dependent on iron imports: Swedish exports rose rapidly after 1650, Russian exports began to dominate after 1750, and by the 1770s England was covering two-thirds of its iron demand with foreign metal (King 2005). An end to the charcoal ceiling on English pig iron production, the elimination of imports, and a massive expansion of iron output (from only about 25,000 t in 1750 to 100 times that much, 2.5 Mt, 100 years later) were possible because of the switch from charcoal to coke. Coke is a lightweight (apparent density of just 0.8–1 g/cm^3) but energy-dense fuel (29 MJ/kg) that is produced from coal by pyrolysis, or heating in the absence of oxygen.

In England, the first use of coke was for drying malt in the 1640s, Shadrach Fox was the first industrialist to use it on a small scale, in a blast furnace, during the 1690s, and, starting in 1709, Abraham Darby became the fuel's best-known promoter (Harris 1988). Coke's initially high production costs limited its rapid adoption, and the two fuels coexisted in England for most of the eighteenth century, and in the United States well into the second half of the nineteenth century. This shift from metallurgical charcoal to coke was part of a much larger energy transition from traditional biomass fuels (woody phytomass, charcoal, and also crop residues, mainly cereal straws) to fossil fuels, first to coals and then to hydrocarbons (Smil 2010b).

This transition introduced fuels that had a generally higher energy content: a kilogram of high-quality coal has nearly twice as much energy (29 MJ) as a kilogram of air-dried wood (17 MJ); a kilogram of liquid fuels refined from crude oil had three times as much energy (42 MJ) as straw used for cooking (14 MJ). The higher energy density also made fossil fuels cheaper to transport and easier to store. But the replacement of charcoal by coke did not bring in a fuel with superior energy content, as both charcoal and coke are essentially pure carbon with a virtually identical energy content of about 30 MJ/kg, but the power densities of their production are orders of magnitude apart.

From Charcoal to Coke

Harvesting coppiced beech or oak would yield, at 5 m^3 or 3.75 t/ha and an energy density (of dry mass) of 19 GJ/t, an annual phytomass harvest of about 0.22 W/m^2. In contrast, a typical late eighteenth-century deep mine—based on detailed data for the Toft Moor Colliery of the 1770s

(Hausman 1980)—would have produced about 15,000 t/year, and all of that coal would have been hauled through a single narrow shaft. With the mine's surface structures occupying no more than a hectare (10,000 m^2) of land, that extraction would have produced fuel with a power density of nearly 1,200 W/m^2, more than 5,000 times higher than the power density of charcoal production.

But more land was needed near the pithead because the fuel was usually sorted at the site to remove associated rocks and to improve the coal's quality. If we assume that the incombustible material (amounting to 10% of the total mass, with a density of 2.5 t/m^3) was deposited nearby in a conical heap just 20 m tall, the area claimed after 50 years of a mining operation would have been no larger than 4,500 m^2, and the overall power density of coal production would have been no less than 800 W/m^2, or 4,000 times higher than the power density associated with harvesting wood.

And we have to make adjustments for conversion to, respectively, charcoal and coke (both calculations disregard the relatively small areas needed for charcoaling or coke ovens). In the early eighteenth century a typical charcoaling ratio (by weight) was five units of wood for a unit of charcoal, which means that in energy terms (with 29 GJ/t of charcoal and 19 GJ/t of wood), it was about 3.3 units of wood for a unit of charcoal, and the power density of charcoal production would have been only about 0.07 W/m^2. Early coking in simple beehive ovens was also inefficient, with up to 1.7 units of coal needed to produce a unit of coke. The overall power density of mid-eighteenth-century English coke production was thus roughly 500 W/m^2, approximately 7,000 times higher than the power density of charcoal production.

This shift from charcoal to coal and the huge difference in accompanying power densities of the two products had many economic, social, and environmental consequences. England could rapidly reduce, and soon eliminate, its dependence on iron imports from Sweden and Russia. The removal of one of the greatest burdens on the country's forests opened the way to reforestation, and the high power density of coal and coke production made it possible to concentrate their output in a progressively smaller number of facilities (large coal mines and coking plants, often attached to large blast furnaces), from which the products could be distributed not only nationwide but also exported abroad.

And the shift made no smaller difference even in those countries that were initially rich in natural forests because they too were not immune to the charcoal limit on iron-making. In the early nineteenth century, American iron-makers had no problem harvesting the needed wood from majestic Appalachian forests, but 100 years later it had become impossible to acquire that much wood from eastern forests. Although smelting in blast furnaces became much more efficient during the course of the nineteenth century (by 1900 it required just 5 kg of wood for every 1 kg of hot metal), the US pig iron output surpassed 25 Mt by 1906, and maintaining that level of production alone (excluding all charcoal needs for further metal processing) would have required (even when assuming a high average increment of 7 t/ha in natural forests) an annual wood harvest of about 180,000 km^2 of forest (Smil 1994).

That is an area the size of the entire states of Missouri or Oklahoma (or one-third of France), equal to a square with a side equivalent to the distance between Philadelphia and Boston, or Paris and Frankfurt—to be harvested annually! Obviously, the power density of producing metallurgical charcoal would have been far too low to enable even forest-rich America to industrialize on that renewable energy basis. And what would charcoal-based iron-making require today? Could the combination of high-yielding clones of fast-growing trees planted in the tropics, higher efficiencies of the best charcoaling techniques, and the lower specific energy requirements of metal smelting provide a land-sparing renewable solution for a modern iron industry?

I address these questions in the book's closing chapter, but anybody even slightly familiar with the advances in modern ferrous metallurgy and with the growth of global pig iron production can anticipate the answer. The combined effect of technical innovations has not brought an order of magnitude reduction in specific charcoal demand (a tonne of fuel needed to produce a tonne of hot metal), while the worldwide iron smelting industry grew by an order of magnitude during the twentieth century and has yet to reach its global peak. This means that the impact of today's charcoal-based iron smelting would be even greater than it would have been more than 100 years ago, when the extent of its practice forced the switch to coke-fueled blast furnaces. Obviously, power densities matter.

2 Quantitative Keys to Understanding Energy

In this chapter I extoll the explanatory utility of rates, sort out the various meanings of power densities used by scientists and engineers, provide a clear definition of the key measure used in this book, offer its brief typology, and explain some of its inherent complications and uncertainties.

By 2014, energy—its resources, consumption, and future supplies, its economic importance and trade and strategic implications, and the environmental impacts of its use—had been a matter of intense public interest, policy-making attention, and expanded scientific inquiries for two generations. The proximate causes of the sudden elevation of energy matters to worldwide prominence during the early 1970s were the quintupling of the price of crude oil sold by the member states of the Organization of the Petroleum Exporting Countries (OPEC) and a temporary embargo of all oil shipments to the United States and the Netherlands (Smil 1987).

As these events were taking place, most people, even in affluent countries, were largely ignorant of basic energy matters. Few could explain the background and the importance of the unfolding changes, and supplying the essential framework would have required assessing realistic options by referring to numbers other than the constantly repeated new record levels of crude oil prices. By 1978 those prices had steadied, but soon they doubled as the Iranian monarchy fell and the ayatollahs took over Iran, at that time the world's fifth largest oil producer. But that price spike was also relatively short-lived, and by 1985 OPEC's oil price had fallen by two-thirds from its 1981 peak. Saddam Hussain's invasion of Kuwait in August 1990 resulted in only a brief price rise, followed by a decade of stable and low oil prices: during the closing years of the twentieth century they were (in constant monies) almost as low as in 1975 (BP 2014).

As concern over high oil prices and the security of energy supply receded, a new concern became prominent with the growing realization of the role played by CO_2 emissions from the combustion of fossil fuels in the process of anthropogenic global warming (IPCC 1995). Understanding planetary energy balance, the physics of greenhouse gases, and the complex atmosphere-hydrosphere-biosphere interactions governing the global biogeochemical carbon cycle is a challenge far more difficult than appreciating the intricacies of global energy supply and use. During the first decade of the twenty-first century—with higher oil prices, fears of resource shortages, and concerns about global warming—anxieties about energy futures were on the rise again, but the quality of the discourse did not improve. People paying attention to post-2000 news heard claims of an imminent peak in global oil production but had no knowledge of crude oil's energy density or of the actual dynamics of the global hydrocarbon reserves.

Similarly, most people who heard that the unfolding global warming would be unprecedented in its rapidity knew nothing about actual CO_2 emission factors or about the relative decarbonization of the global energy supply. Understanding complex energy matters, formulating informed arguments, and making sensible choices can be done only on the basis of a quantitative understanding that must be both relatively broad and sufficiently deep. There is a natural progression in this understanding, from simple quantities to rates that relate those variables to basic physical attributes, to time and space.

Power of Rates

Most phenomena are best understood when expressed as rates, specific ratios relating two variables. In the scientific sense, all rates are derived quantities. They are defined in terms of the seven base units of the Système international d'unités (SI): length (the SI base unit is a meter, m), mass (kilogram, kg), time (second, s), electric current (ampere, A), temperature (kelvin, K), amount of substance (mole, mol), and luminous intensity (candela, cd). Speed (velocity, m/s) is perhaps the most commonly used rate in everyday affairs, while the rates frequently encountered in scientific and technical inquiries include mass density (kg/m^3), amount of substance (mol/m^3), and luminance (cd/m^2)—as well as energy and power.

Those energy-related derivations start with force (newton, N, is m × kg/s^2). The energy unit, joule (J); is a newton-meter (m^2 × kg/s^2), and the unit of power (watt, W) is simply the rate of energy flow (J/s or m^2 × kg/s^3). In turn, these units can be used in specific rates as they are related to the base variables of length, mass, time, substance, and current, or to individuals or groups of people, to give fundamental insight into the nature and dynamics of energy systems: only when the absolute values are seen in relative terms can we truly appreciate their import and make revealing historical and international comparisons. Tables in the appendix list all principal units and their multiples and submultiples.

Certainly the most common class of these higher-order derivatives comprises quantities prorated for an individual in a given data set, and average national per capita rates are perhaps the most frequent use of this measure. When they are used to quantify natural endowment (water resources, cropland, standing forest phytomass, fossil fuel reserves) they refer to a particular year, but when they are used to express average supply or consumption they become double rates, prorated not only per capita but also over a specific time period. In the case of food supply (measured in kcal/capita or in MJ/capita) the rate is per day, and—as illustrated by contrasting the United States with Japan—those rates alone tell us a great deal about a nation's food supply, dietary habits, overeating, and excessive food waste.

US food per capita availability (the total at the retail level) now averages about 3,700 kcal/capita/day, and as that mean places babies and octogenarians in equivalency with adults (whereas the normal daily food intake of babies and octogenarians should be either below or barely above 1,100 kcal), it implies a supply of more than 4,000 kcal/day for adults (FAO 2014; USDA 2013b). Obviously, if that were an actual average consumption, the American population would be even more obese than it already is. US food consumption surveys show an actual daily intake averaging only about 2,000 kcal/capita. These surveys are based on individuals' recall of food intake in a day and hence are not highly accurate, but even after adding 10% to their mean there is still a gap of 1,500 kcal/day, which means that the United States wastes 40% of its food supply.

An excellent confirmation of this loss comes from the modeling of metabolic and activity requirements of the US population by Hall and co-workers (2009). They found that between 1974 and 2003, that rate was between 2,100 and 2,300 kcal/day, while the average food supply rose from about

3,000 to 3,700 kcal/day, resulting in food waste rising from 28% of the retail supply in 1974 to about 40% by 2004. In contrast, no other affluent economy has been wasting so little food as Japan: the country's recent average per capita food supply has been only between 2,500 and 2,600 kcal/day, while annual studies of dietary intake show consumption of just over 1,800 kcal/capita, resulting in food waste of less than 30% (Smil and Kobayashi 2012).

In the case of raw materials and finished products, and for such key financial indicators as GDP or disposable income, per capita rates are usually given for a calendar year, while the availability of such essential quality-of-life indicators as number of doctors or hospital beds is expressed, obviously, per 1,000 people rather than per capita. But all of these indicators also illustrate a common problem with average per capita rates: their simplistic international comparisons—ignoring differences in the quality of statistics and, even more important, in the qualitative differences and the wider socioeconomic setting of specific variables—may mislead and confuse rather than reveal and explain. Similar caveats apply, to a greater or lesser extent, even to seemingly straightforward energy-related variables.

Population Density
There are three kinds of densities in common use. The first and the most commonly used rate relates the number of individual items (be they organisms, people, or artifacts) to a specific area. This spatial density is not among the SI derived units, but the measure is common in ecological and population studies. In the first case, the densities of small organisms (plants, insects, invertebrates, small mammals) are measured as the number of individuals or their collective mass per square meter (m^2). A more common rate for larger animals is individuals or total mass per hectare (ha) or square kilometer (km^2), a rate that is also used for numbers and mass of trees, and for other phytomass (woodlands, grasslands, wetlands).

When population densities are expressed in relation to agricultural land, the rate is almost always given as an inverted value, ha/person, ranging from about 1.25 ha/capita in Canada and 0.5 ha/capita in the United States to less than 0.1 ha/capita in China and Bangladesh and to negligible areas (even to nothing) in many small island and desert nations (FAO 2014). In agronomic studies the density of planting or transplanting is also expressed per hectare. For example, corn densities in Iowa are now as high as 100,000 plants/ha, but while yields initially increase with higher densities, they level off, and optimum yields with no soil quality constraints range between 70,000 and 80,000 plants/ha (Farnham 2001). For comparison, the highest-yielding hybrid rice varieties are grown with up to 200,000 plants/ha (Lin et al. 2009).

Regional and national population densities are measured in individuals/ km^2. This measure is particularly misleading as it prorates populations over entire national territories regardless the land's habitability: even in countries with a relatively homogeneous population density it will hide large regional differences. For example, even in the generally densely populated Netherlands it subsumes differences ranging over two orders of magnitude, from cities with more than 5,000 people/km^2 to rural regions with fewer than 50 people/km^2. Much larger differences in intranational population densities are common in all countries with very large territories, as well as in nations located in arid and semiarid climates, particularly in North Africa. Canada is an excellent example of these huge disparities among affluent countries; Egypt, with less than 4% of its territory under annual and permanent crops, offers the extreme example in the arid world. Moreover, continuing urbanization means that nationwide population densities are becoming universally less representative.

The density of human populations can be also expressed in mass terms, and in this case care must be taken to specify the mass: either as live weight or in absolutely dry terms. The most densely populated parts of Asia's still expanding megacities have residential densities on the order of 50,000 people/km^2, which (using a conservative age- and sex-weighted mean of 45 kg/capita) translates to a live weight anthropomass of more than 2 kg/m^2, a rate unmatched by any other mammal and three orders of magnitude higher than the peak seasonal zoomass of large herbivorous ungulates grazing on Africa's richest grasslands (Smil 2013a).

The average densities of services (be they health, commercial, or recreational: doctors' offices, hospitals, food stores, children's playgrounds, sports playing fields) offer simple but revealing ways to focus on spatial inequities and their consequences. For example, a study by Bonanno and Goetz (2012) revealed that even after controlling for missing variables, biases, and lags, the density of stores selling fruits and vegetables (as opposed to many small establishments where only packaged and fast foods are available) was associated with higher shares of adults who consumed fruits and vegetables regularly and had lower obesity rates.

Mass and Energy Density

The second density category includes those derived SI units that relate variables to volume: mass density and energy density. SI mass density is measured in kg/m^3, but in practice it is often expressed also in g/cm^3, kg/dm^3, or t/m^3 (the number will be identical for these three rates). The densities of common materials (all expressed in g/cm^3, with water as the yardstick at 1) range from 0.65 to 0.75 for most wood species to just short of 1 for plastics (polyethylene goes from 0.915 to 0.970 for its low- and high-density varieties). Concrete has densities between 2.2 and 2.4, aluminum and its alloys are just above that, at 2.6–2.7, and steel alloys cluster mostly between 7.7 and 7.8 (Smil 2013b).

In SI terms, energy density is a derived unit measured in J/m^3, but in energy publications this density is often expressed (with the exception of gases) in mass terms as MJ/kg or GJ/t. This may be a cause for confusion because in SI nomenclature, J/kg is a derived unit called specific energy. In SI units this specific energy is also often measured in J/g, MJ/kg, or GJ/t. Accurate conversions between these two rates (from volume to mass or, in SI terms, from energy density to specific energy) require analyses of individual fuels. Energy density is one of the key determinants of the structure and dynamics of an energy system: there are many reasons to prefer sources of high energy density, particularly in modern societies demanding large and incessant flows of fuels and electricity.

Obviously, the higher the density of an energy resource, the lower are its transportation (as well as storage) costs, and this means that its production can take place farther away from the centers of demand. Crude oil has, at ambient pressure and temperature, the highest energy density of all fossil fuels (42 GJ/t), and hence it is a truly global resource, with production ranging from the Arctic coasts to equatorial forests and hot deserts, and with enormous investment in an unmatched worldwide shipping infrastructure (long-distance pipelines, oil loading and offloading terminals, giant tankers) and high-throughput processing in large refineries.

In contrast, wood and crop residues (mostly cereal straws), the two most common traditional phytomass fuels, have low energy densities, with crop residues at just 14–15 MJ/kg, while wood (depending on the species and the degree of dryness) ranges from less than 15 MJ/kg for fresh-cut branches to about 17 MJ/kg for air-dried wood and to almost 20 MJ/kg for absolutely dry woody matter. Charcoal is the great exception as the pyrolysis of wood produces nearly pure carbon with a specific energy of nearly 30 MJ/kg. This means that twice as much straw or air-dried fuelwood has to be burned to yield the same amount of energy, and charcoal's smokeless combustion—as opposed to the often very smoky burning of wood in open fires or in poorly designed stoves, which causes millions of premature deaths every year (Subramanian 2014)—is another welcome advantage of charcoal for indoor use. Traditional societies paid a high energy and environmental price for these advantages because making a unit of charcoal in simple clay kilns required up to 10 units of wood.

Some coal varieties are as energy dense as charcoal: early coal extraction often produced the highest-quality anthracites (much like charcoal, they are nearly pure carbon, with a specific energy of up to 30 MJ/kg) and excellent bituminous coal (25–27 MJ/kg). As coal mining progressed, the average energy content of the product declined, particularly with the shift to the less expensive and much safer surface extraction methods. The specific energies of most steam coals are now 22–26 MJ/kg, but those of the poorest lignites are less than 10 MJ/kg. The transition to liquid hydrocarbon introduced fuels of unrivaled energy density: crude oils range from gasoline-like light liquids to heavy varieties that might need heating for transportation, but their specific energies span a narrow range around 42 MJ/kg; with densities ranging between 0.75 and 0.85 kg/m^3, this translates to 32–39 MJ/m^3. Because of these high densities, refined liquid fuels dominate all transportation (and other portable) uses.

Natural gas (mostly methane, CH_4 with small shares of higher alkanes, collectively known as natural gas liquids) has a higher hydrogen share (75%) than liquid hydrocarbons and hence it contains 53.6 MJ/kg—but its liquefaction requires considerable energy input for refrigeration, and it is used only for (still expensive) intercontinental shipments of liquefied natural gas (LNG). In its gaseous form methane's energy density is 35 MJ/m^3, amounting to less than 1/1,000 the energy density of gasoline. A higher hydrogen content explains this progression of higher energy density. The very low energy density of natural gas is no problem when the fuel is delivered by pipelines for stationary combustion in electricity-generating gas turbines or in industrial, commercial, and household furnaces.

Finally, in the third category of density are those derived SI rates that relate a basic quantity (current, luminous intensity) or a derived unit to space: they include current density (A/m^2), luminance density (cd/m^2), illuminance density (lm/m^2), electric flux density (C/m^2), and magnetic flux density (Wb/m^2). Power density is not an official name given to any derived quantity on the list of SI units. The rate, measured in W/m^2, is listed among the derived quantities but is given a rather restricted scope: it is called heat flux density or irradiance (flux of solar or other radiation per unit area). *In this book, power density always refers to the quotient of power and land area, and I demonstrate that this rate is a key variable in energy analysis because it can be used to assess the suitability and potential of specific energy resources, the performance and operating modes of energy converters, and the requirements and structures of complex energy systems.*

In all of these respects, energy density is an insufficiently revealing measure. For example, crude oils, regardless of their appearance and physical differences, have a uniformly high energy density, but if they were present in minuscule reservoirs strewn over a large area, the power density of their extraction would be too low to warrant their commercial exploitation; in contrast, the Middle Eastern oil fields supply the whole world precisely because they produce high-energy-density fuel with unmatched power densities. Similarly, US bituminous coal, mined with high power densities, is shipped to Europe, where energy-dense hard coal deposits remain unexploited because their thin seams deep underground cannot be extracted at acceptable cost (or with power densities comparable to American mining).

Power Density: Sorting Out the Rates

There is no single, binding, universal definition of power density as different science fields and different branches of engineering—including electrochemistry, telecommunications, and nuclear electricity generation—have used the term for a variety of kindred but distinct rates, with mass, volume, and area as denominators. Then there is a revealing, and virtually universal, notion of power density as the energy flux in a material medium, a concept that can be used to assess the potential performance of all modern commercial energy conversions. And in this book I relate energy flux to its fundamental spatial dimension by quantifying power that is received and converted (or that is potentially convertible) per unit of land, or water, surface. Before I begin doing so I will briefly review the other applications of power densities.

In electrochemistry, power densities, expressed both in volume (W/cm^3) and in mass (W/g) terms, are used to rate the performance of batteries. While energy density (J/g) measures the specific energy a battery can hold (a higher rate implying, obviously, a longer run time), power density measures the maximum energy flux that can be delivered on demand in the short bursts of electricity required for tools, medical devices, and in transportation. Early lead acid batteries delivered less than 50 W/kg, but by the end of the twentieth century the power densities of these massively deployed (above all in cars) units were between 150 and 300 W/kg. Utility-grade batteries can deliver up to 40 MWh of electricity with efficiencies of up to 80%, and their power densities range between 200 and 400 W/kg at an 80% charge level. Since the year 2000 the best performance of experimental lead acid batteries has been boosted by the addition of high-surface-area carbon: its addition adds only up to 3% of weight but increases surface area by more than 80%, to yield a power density of more than 500 W/kg; the ultimate target is 800 W/kg (Svenson 2011).

More expensive nickel-cadmium batteries can supply up to about 1 kW/kg, and the increasingly common lithium-ion batteries deliver well in excess of that (Omar et al. 2012; Rosenkranz, Köhler, and Liska 2011). High-energy Li-ion batteries deliver more than 160 Wh/kg, but their power density is less than 10 W/kg; in contrast, very high-power Li-ion batteries, commonly used in electric vehicles (where they have to deliver more than 40 kW), rate less than 80 Wh/kg but have power densities up to 2.4 kW/kg at an 80% charge; as the density of these batteries is about 2.1 kg/L, that translates to almost 5 kW/L. By 2014 the best commercially available rates were about 2.8 kW/kg, and the world's highest-power-density Li-ion car battery, a prismatic cell revealed by Hitachi in 2009, can discharge as much as 4.5 kW/kg (Hitachi 2009).

In nuclear engineering, the average core power density is the amount of energy generated by the specific volume of the reactor core, a quotient of the rated thermal reactor power and the volume of the core (IAEA 2007). Its value is usually expressed in kW/dm^3 (kW/L), and the range for all reactors listed in the International Atomic Energy Agency's Prius Database is

1–150 kW/dm^3. Early British Magnox reactors and advanced gas reactors (AGRs)—designs that used graphite moderator and CO_2 cooling when they pioneered the UK's commercial fission electricity generation during the late 1950s—had a power density of, respectively, 0.9 and 3 kW/dm^3. The rates for pressurized water reactors (PWRs, water-cooled and water-moderated), the dominant choice for commercial nuclear electricity generation around the world, are mostly between 70 and 110 kW/dm^3. The highest rates (in excess of 700 kW/dm^3) were reached in experimental fast breeder reactors cooled by molten salt (Zebroski and Levenson 1976; IAEA 2007). Power densities in traveling-wave reactors, now under development, would be about 200 kW/dm^3 within the active fission zone.

Telecommunication engineers routinely calculate the power density for energy received from transmissions emanating from both isotropic and directional antennas. One of the main reasons for this is to make sure that human exposures to nonionizing radiation do not exceed accepted safety standards. The US Federal Communications Commission (FCC) set the maximum permissible exposure for power density for transmitters operating at frequencies between 300 kHz and 100 GHz (FCC 1996). Between 30 and 300 MHz (very high-frequency wavelengths of 1–10 m that carry FM radio and television broadcasts) the limit is 1 mW/cm^2 for occupational exposures averaging 6 minutes, and just 0.2 mW/cm^2 for the general population and uncontrolled exposure averaging 30 minutes.

For shortwave broadcasts (frequencies of 2.3–26.1 MHz), the general population limit in mW/cm^2 is 180/f^2: this means that an international BBC broadcast at 15 MHz would allow for the maximum exposure of 0.8 mW/cm^2. Typical shortwave radio transmitters have a power of 50–500 kW, while many longwave transmitters rate more than 500 kW, and the world's most powerful one requires 2.5 MW. The power density (P_D) of an isotropic antenna (radiating energy equally in all directions) is simply a quotient of the transmitted power (P_t, peak or average) and the surface area of a sphere at a given distance: $P_D = P_t/4\pi r^2$. A 100-kW transmitter would thus produce a P_D of 0.8 nW/m^2 at a distance of 1,000 km, equal to only one-millionth of the allowable exposure.

In reality, most radio antennas have considerable transmission gain (G_t), which is created by suppressing upward and downward directions and concentrating the output toward the horizontal plane. After correcting for this intervention ($P_D = P_t G_t/4\pi r^2$), a 100-kW shortwave transmitter with a gain factor of 10 will have an effective radiated power of 1 MW and a P_D of 8 nW/m^2 at 1,000 km. Shortwave broadcasting antennas rely on particularly narrow beam widths in order to transmit their signal between continents; so do, of course, radar antennas that require a high gain of up to 30 or 40 dB (that is, a G_t between 1,000 and 10,000) in order to pinpoint distant targets (Radartutorial 2014).

For the ultra-high frequencies (1.9 GHz) used by cell phones, the general population limit is 1 mW/cm^2, while the actual received maxima near a cell phone tower are 10 µW/cm^2, that is, just 0.01 mW/cm^2. Higher power density exposures apply to time-varying electric, magnetic, and electromagnetic fields between 30 and 300 GHz; these wavelengths of 1–10 mm are the highest radio frequency just below the infrared radiation. The International Commission on Non-Ionizing Radiation Protection puts the maximum power densities at 50 W/m^2 for occupational exposure and 10 W/m^2 for general public exposures (ICNIRP 1998).

Umov-Poynting Vector

Electrochemists, reactor physicists, and radio engineers thus use power densities with three different denominators (mass, volume, and area), but a universal approach makes it possible to assess the power density of virtually all energy conversions by quantifying the energy flux per unit area of the converter's surface. In 1884 John Henry Poynting (1852–1914; fig. 2.1), a professor of physics at the University of Birmingham, set out to prove

that there is a general law for the transfer of energy, according to which it moves at any point perpendicularly to the plane containing the lines of electric force and magnetic force, and that the amount crossing unit of area per second of this plane is equal to the product of the intensities of the two forces, multiplied by the sine of the angle between them, divided by 4 π while the direction of flow of energy is that in which a right-handed screw would move if turned round from the positive direction of the electromotive to the positive direction of the magnetic intensity. After the investigation of the general law several applications will be given to show how the energy moves in the neighbourhood of various current-bearing circuits. (Poynting 1884, 344)

This directional transfer of energy per unit area (energy flux density, measured in W/m^2) became known as the Poynting vector, but, as is often the case in scientific discovery, a Russian physicist, Nikolai Alekseevich Umov (1846–1915; fig. 2.1), formulated the same concept a decade earlier

Figure 2.1
John Henry Poynting (left) and Nikolay Alekseevich Umov, the two originators of
the concept of energy flux density. Stipple portraits by Noli Novak.

(Umov 1874), and hence in Russia the measure has been known as the
Umov-Poynting vector. Piotr Leonidovich Kapitsa (1894–1984), Ernest
Rutherford's student and Nobel Prize winner in physics (in 1978, for his
basic discoveries in the area of low-temperature physics), pointed out that
the vector can be used to assess all energy conversions to reveal the "particular restrictions of these various flows," which are often ignored, resulting "in wasting money on projects that can promise nothing in the future"
(Kapitsa 1976, 10).

The Umov-Poynting vector thus offers a fundamental assessment of
energy converters in all cases where

the density of the energy influx is limited by the physical properties of the medium
through which it flows. The rate at which energy can be made to flow in a material medium is restricted by the velocity (v) of propagation of some disturbance (a
mechanical wave or heat flow, for example) and the energy density (U) of the disturbance. The rate of flow (W) is always in a particular direction (it is a vector, like an
arrow). Vector W is equal to vector v times U and proves very convenient for study-
ing processes of energy transformations. (Kapitsa 1976, 10)

Of course, the final value must be multiplied by appropriate factors in
order to account for maximum efficiencies. This is well illustrated by

Box 2.1
Maximum power of a wind turbine

$$P = 1/2\rho A v^3$$
$$= 0.5 \times 1.2 \times 7{,}854 \times (12)^3$$
$$= 4712.4 \times 1728$$
$$= 8.14 \text{ MW}$$

looking at electricity generation by a large modern wind turbine. Its 50-m-long blades will sweep an area A of roughly 7,854 m^2; with a wind speed v of 12 m/s and an air density (at 20°C) of 1.2 kg/m^3, the kinetic energy density ($U = 0.5mv^3$) will be 1,037 J/m^3, and the maximum power of the machine would be 8.14 MW.

This result must be corrected for the theoretical maximum efficiency: Betz (1926) established that its limit is 16/27 (0.59) of the potential. Multiplying 8.14 MW by 0.59 sets the maximum turbine power at 4.8 MW, but the actual performance is considerably lower because of unavoidable energy losses (in gearing, bearings), and the correction factors range between 0.35 and 0.45 even for the best-designed modern wind turbines. I chose the blade radius of 50 m because it matches that of the GE 2.5-MW series turbine (GE Power & Water 2010; fig. 2.2). The actual power coefficient of this large machine is only 0.3, and the effective power density of its electricity generation is 310.9 W/m^2 of the area swept by its blades.

The vector can be used to find the limits of electric generators or combustion engines and, as Kapitsa noted at the outset of his paper, to refute some apparently appealing proposals: he related how he was asked to disprove the idea suggested by his teacher, the famous physicist Abram Fedorovich Ioffe, to use electrostatic rather than electromagnetic generators for large-scale electricity production. Electrostatic generators would be easier to build and could feed high voltage directly into the electricity grid. But to avoid sparking, the electrostatic field is restricted by air's dielectric strength, and to generate 100 MW (a rate sufficient to supply the electricity needs of nearly 80,000 average American consumers), the electrostatic rotor would need to have an area of about 400,000 m^2 (nearly 0.5 km^2), obviously an impossible requirement.

Figure 2.2
GE 2.5-MW wind turbines. Photo reproduced by permission.

The maximum power that can be transmitted during combustion from a burning medium to the working surface (an engine piston or rotating turbine blades) is the product of gas pressure, the square root of its temperature, and a constant dependent on the molecular composition of the gas. The vector also makes it clear why some very efficient energy conversions are not suitable for high-power supply because of the low power densities. Fuel cells are an excellent example of such limitations: their peak theoretical efficiency of transforming chemical energy into electricity is about 83%, but low diffusion rates in electrolytes limit their power density to about 1 W/cm^2 of the electrode.

This means that the working surface of fuel cells delivering 1 GW (a rate easily needed by a large city) would have to be on the order of 100,000 m^2. Obviously, the power density of fuel cells is too low to provide the centralized base-load supply for modern urban, high-energy settings (Brandon and Thompsett 2005). In contrast, in modern large thermal turbogenerators rated at 1 GW (enough to provide electricity for at least 750,000 average US consumers) the high velocities and high temperatures of the working medium (steam superheated to 600°C, traveling at 100 m/s with a density of 87.4 kg/m^3 at 30 MPa) create power densities as high as 275 MW/m^2

across the area swept by the longest set of blades rotating at 3,600 rpm. I return to Umov-Poynting vector densities when reviewing the performances of some modern energy converters.

My final example of the prevailing lack of consensus regarding the application of power density can be found by consulting Elsevier's six-volume *Encyclopedia of Energy* (Cleveland et al. 2004). Four authors use the measure in four different ways. In the first volume, Michael M. Thackeray (2004, 127), reviewing batteries for transportation, defines the rate as "power per unit of volume, usually expressed in in watts per liter." In the third volume, J. M. German (2004, 197), writing about hybrid electric vehicles, employs mass denominator, "the power delivered per unit of weight" (W/kg). I use it, also in the third volume, when looking at land requirements of energy systems (Smil 2004b, 613), as "average long-term power flux per unit of land area, usually expressed in watts per squared meter." And in the sixth volume, Arnulf Grübler (2004, 163), writing about transitions in energy use, defines it as the "amount of energy harnessed, transformed or used per unit area."

W/m^2

As already noted, my primary measure of power density in this book is not the energy flux per unit of the working surface (vertical or horizontal) of a converter (as in the Umov-Poynting vector) but per unit of the Earth's surface, for comparability always expressed in W/m^2. I have been advocating the use of this measure since the early 1980s (Smil 1984) and chose it as key analytical variable in my first synthesis of general energetics (Smil 1991) and again in its thoroughly revised and expanded sequel (Smil 2008). Since the late 1990s, power density, expressed as energy flux per time per unit of horizontal surface, has been receiving more attention as a result of the growing interest in renewable energy resources and their commercial conversions to fuels and electricity. The power densities realized by harnessing these energy flows are appreciably lower than the power densities of fossil fuel–based systems, something that even often uncritical proponents of wind and solar energy—the new renewables, in addition to the two traditional renewable energies, hydro energy and fuelwood—cannot ignore.

Perhaps the greatest advantage of power density is its universal applicability: the rate can be used to evaluate and compare all energy fluxes in nature and in any society. In this book I use it for a systematic assessment of all important energies, be they natural (renewable) flows or fossil fuels burned to produce heat for many direct uses or as the means of generating thermal electricity. Because fossil fuels are the dominant source of primary energy, I look first at the power densities associated with producing, processing, and transporting coal, crude oil, and natural gas, and then at combustion processes in general and at the thermal generation of electricity in particular. A significant share of the latter activity is also energized by uranium fission, and I quantify the power densities of the entire nuclear fuel cycle.

I also quantify, in a brief historical survey and in more detailed sectoral comparisons, the hierarchy of power densities of energy uses and the challenges of heat rejection arising from highly concentrated energy conversions. In the penultimate chapter I summarize the book's main findings to demonstrate why power densities matter and why their appreciation and assessments should be among the key ingredients of any attempts to understand the energy predicament of modern, high-energy civilization and to guide the development of its future energy supplies. Although there is no imminent global prospect of running out of fossil fuels (the very notion of running out as a complete physical absence of a resource is incorrect, as rising prices will end an extraction long before the last reserves are exhausted), such resources are finite, and eventually our civilization will have to make the transition to renewable energy sources.

The early stages of this transition have been unfolding during the past generation, and in the book's last chapter I explain some of the challenges involved in shifting from a high-power-density system of energy supply based largely on fossil fuels to low-power-density renewable energy flows and their commercial energy conversions to produce electricity as well as substitutes for oil-based liquid fuels. But before starting a systematic review of specific power densities I should take a closer look at the qualitative differences that are hidden by a simple quantitative expression. They affect the rate's numerator: all energies can be converted to a common denominator, but that measure cannot convey their qualities. And the qualities are no less important as far as the denominator is concerned: space attributable to energy production and use may be easily measured in common units, but its qualitative attributes range across a very wide spectrum of land cover transformations.

Typology and Caveats

Power density is an inherent property of all forms of energy production and energy use. These processes unfold on scales ranging from subatomic to universal, but we are interested above all in their varied and ubiquitous commercial manifestations that sustain our civilization. Power density, however, is not a primary design parameter. The energy systems of modern societies are designed for overall performance and reliability achieved at acceptable cost: annual output (and hence maximum available power), price per unit of product, and desired parameters of safety and environmental impacts govern their construction and performance, and a specific power density is simply a key consequence of their operation.

The measure is deceptively simple—power (energy flow per unit of time) per unit of area, in standard form expressed in W/m^2—but both the numerator and the denominator hide important differences and complexities. Neither of them can convey the quality of the two measured variables: watts and square meters are measures that will accurately quantify power and area without telling us directly anything either about the quality of a specific energy flow (its environmental externalities, its ease of use, its reliability, its cost, its durability) or about the initial quality of land that was claimed by energy production and use and about the eventual extent of its destruction or alteration. A closer look at some of the most important qualitative attributes of the two variables is important in order to appreciate the inherent deficiencies of the measure.

This information loss is a common problem when using single measures for variables whose quality matters at least as much as, or more than, their quantity. Food consumption is an excellent example of this information deficiency. Every foodstuff can be quantified in terms of energy density (kcal/g, MJ/kg), and average per capita food intake can be expressed in kcal/day or J/day, but these commonly used quantities hide fundamental nutritional qualities. They tell us nothing about the shares of macronutrients (carbohydrates, lipids, proteins) in foodstuffs, about their specific qualities (all animal proteins are nutritionally superior to plant proteins), or about the presence of micronutrients. As a result, people can consume energetically adequate quantities of food but be malnourished; unfortunately, that is not an uncommon situation for some population groups even in many affluent countries, particularly as far as some key micronutrients (above all iron and zinc) are concerned (Smil 2013c).

Energy Qualities

Quantification of the numerator is subject to frequent but minor inconsistencies arising from the conversion of fuels to common energy equivalents. The most common cause is the slightly different assumptions made regarding average energy densities (J/kg or J/m^3) of fuels and the choice (often undefined) of using either higher or lower heating values. But by far the greatest challenge arises from converting various forms of electricity to a primary energy equivalent. With electricity generated by fuel combustion, it is simply a matter of adjusting for the overall efficiency of thermal generation: an aging coal-fired station may have an efficiency of just 33%, the latest model may convert 42% of coal into electricity, and for combined-gas cycle generation the rate may be as high as 60%. Consequently, the primary energy equivalents of the three processes will be respectively 10.9, 8.6, and 6.0 MJ/kWh.

But there is no universally valid (or accepted) way of converting primary electricity that includes nuclear, solar, wind, geothermal, and hydro generation. The two commonly used procedures differ by roughly a factor of three: the simple thermal equivalent is 3.6 MJ/kWh; conversions using the prevailing efficiency in large central fossil-fueled thermal stations range from 8.6 to 11.0 MJ/kWh. This disparity explains most of the differences in primary energy totals offered by data aggregators and statistical services. For example, British Petroleum's widely quoted *Statistical Review of World Energy* uses the straight thermal equivalent (BP 2014), while the International Energy Agency uses that equivalent (3.6 MJ/kWh) for all forms of renewable-source electricity generation but calculates the primary energy equivalent of nuclear electricity by assuming that 1 kWh equals 10.9 MJ (IEA 2014). These realities also mean that no matter which conversion rates are chosen, the power densities of fuel combustion are always qualitatively different from the power densities of primary electricity use.

The obvious finite versus renewable energy source dichotomy aside, the key difference, and one with enormous environmental and hence socioeconomic implications, is that fossil fuel combustion is always associated with CO_2 emissions, the principal source of anthropogenic carbon releases and the leading cause of human interference in the global biogeochemical carbon cycle. The combustion of coals and hydrocarbons also results in the emission of oxides of nitrogen (precursors of photochemical smog), and the burning of coal and of many liquid fuels releases sulfur oxides (key contributors to acid deposition). In contrast, primary electricity (from nuclear fission and renewable energy flows) is not a direct source of CO_2 emissions, but carbon emissions do, of course, result from the construction and maintenance of the requisite facilities and converters. Other qualitative differences pertain to the convenience and flexibility of energy use.

Electricity is always easier to use (with a flip or push of a switch) than fuels, although in household settings a modern high-efficiency natural gas furnace with electronic ignition connected to a programmable thermostat comes close in reliable convenience: except for an annual checkup, its operation is entirely automatic. And the only two important commercial applications in which electricity cannot compete with fuel are (as already explained) the production of iron from its ores (where coke or charcoal remains indispensable) and commercial flying, where kerosene remains the only practical choice for jetliners. The capability of experimental unmanned solar-powered airplanes is orders of magnitude behind (Paur 2013). These planes can stay aloft for many hours, and *Solar Impulse* crossed the United States—but it needed 12,000 photovoltaic (PV) cells to do so, it carried only its pilot, its speed averaged less than 50 km/h, and its wingspan was almost the same as that of a wide-body Boeing 747–400, which can carry more than 500 people at the speed of Mach 0.92.

Space Qualities

The denominator of power densities—a unit of horizontal surface—is seemingly a much simpler measure, but a brief reflection makes it obvious that this ordinary quantity subsumes some very different spatial and functional qualities. For all photosynthetic production in fields, grasslands, and forests the denominator is simply a unit of land covered by cultivated or wild plant species, with a yield expressed in W/m^2 rather than (as in agriculture) in t/ha or (in forestry) in m^3/ha. For fossil fuels it is the unit of surface disturbed by extraction (coal mining activities, oil and gas drilling) or otherwise claimed by requisite infrastructures (roads, rights-of-way for pipelines). For thermal electricity-generating plants it is a combination of permanent structures (boiler and generator halls, other buildings, cooling towers) and necessary infrastructures, be they on-site (in coal-fired plants mainly coal storage, switchyard, fly ash depository) or off-site (ROWs for high-voltage transmission lines). For nuclear stations it also includes also the exclusion (safety) zone surrounding a plant.

For renewable energy conversions it gets somewhat more complicated. For the direct combustion of biofuels it is obviously just the area of harvested phytomass: for liquid biofuels it is the cultivated land used to grow plant feedstocks that are converted to ethanol or biodiesel. But the attribution becomes uncertain when crop or logging residues are used to produce cellulosic ethanol: they are by-products of crop cultivation or wood cutting and would be produced even if they were not destined for enzymatic fermentation. For hydroelectricity it is the entire surface of a reservoir created by a dam, but there are several possibilities to consider, including the maximum reservoir design level (which may not be attained for years after dam completion), the average level, determined as the mean of seasonal fluctuations, or the modal level that prevails for most of the year. In any case, for all large projects, land claimed by the dam and associated infrastructure amounts to only a small fraction of the area inundated and periodically partially reexposed by the extreme reservoir levels.

To be comparable with other conversions, the power densities of wind-generated electricity require a transposition from vertical to horizontal planes. The working surfaces of wind turbines are vertical, but the power density of wind generation is calculated as a ratio of electricity produced per unit of horizontal land surface. The power densities for solar PV electricity generation (and also for solar water heating) should be expressed, for conformity's sake, per unit of horizontal surface area, but they are usually given per unit area of actual working surface (on the ground or on a roof), which is fixed at an angle to optimize radiation capture or (a much more expensive solution) which uses automated tracking to maximize the exposure.

As a result, power densities are calculated by using several obviously distinct types of land covers or land uses, and the similarities (often identities) and differences embodied in these rates cut across production modes, infrastructural arrangements, and diverse energy uses. My attempt at a typology of space that becomes the denominator in calculating power densities offers a few fairly simply classifications based on obvious structural and functional attributes: I consider the degree of transformation, project longevity, the likelihood that the land claimed by energy facilities will regain its former function (or at least some of it), and the possibility of concurrent uses of land (or water) that is devoted to the production or conversion of energies.

Space claims by modern energy conversions are hierarchical. On the most intensive end of the spectrum is the obliteration extreme, whereby not only the original plant cover but even the physical appearance of natural surfaces has been entirely erased and replaced by structures that are completely and exclusively devoted to a site's new function, be it extraction, transportation, processing, or conversion. Then comes an extended continuum of impacts of diminishing intensity that eventually joins the other extreme, land that is claimed, owned, or managed by energy industries but whose surfaces have retained their previous (and often entirely undisturbed) soil and plant cover.

The most diverse group in the first category comprises the structures required by modern energy extraction, transportation, and processing and by electricity generation. Every industry has many structures and assemblies of this kind, some highly standardized, others of specific design: buildings that house hoisting machinery, sorting and washing facilities, and storage silos for underground collieries; drill pads, wellheads, gas processing, liquefaction, and regasification plants, compressor stations, refineries, and loading docks for the oil and gas industry; boiler and turbogenerator halls, electrostatic precipitators, desulfurization units, and cooling towers for thermal power plants. In light of the high power densities of fossil fuel combustion and nuclear fission, it is not surprising that most of the structures housing these activities account for only a small fraction of total land claims by thermal power plants.

In addition to buildings that house machinery and processing facilities there are also office buildings, and many energy extraction and conversion sites must have infrastructural components whose construction and operation result in destruction of original soils and plants: these include often extensive areas of impervious surfaces (roads and walkways, large parking lots, storage sheds or open lots with stacked components and parts, railroad yards, switchyards) and frequently even more extensive fuel and waste depositories (large coal yards, oil and gas tanks, ponds storing captured fly ash and sulfates produced by flue-gas desulfurization).

Relatively large shares of areas claimed by energy projects have retained their soils and vegetation but have been fragmented to such an extent that they have lost the capability to provide many of their former ecosystemic services. Satellite images illustrate a wide range of these energy-related fragmentations. The development of oil and gas fields on grasslands or in forests requires the construction of access roads and the drilling of many, often fairly closely spaced, wells; drilling pads and wellheads create pockmarked landscapes (often in a regular, gridlike manner) that are dissected by roads, and further disturbances are created by installing pumps, compressors, and (aboveground or buried) gathering pipelines. Measurements will show that all aboveground structures claim only a small share of the affected area, but while most of the land is undisturbed, its fragmentation excludes any commercial use and degrades its value as habitat for plants and animals. Similar pockmark fragmentation can be also seen (though usually on a smaller scale) with in situ leaching of uranium.

Large-scale PV electricity generation with massed rows of panels fastened to elevated steel supports entails much greater interventions. Support columns, arrayed in regular formations, disturb soil, panels shade the ground, and space must left between the rows for access needed for maintenance and regular cleaning. In contrast, large wind farms can be seen as perhaps the least disturbing example of this fragmentation. As I explain in some detail in the next chapter, turbine siting requires some minimum spacing between adjacent towers, and this distance increases with machine capacity.

Consequently, large modern wind turbines stand hundreds of meters apart, and even when the area of all access roads and transformer stations is added to the space occupied by their concrete foundations, more than 95% or even more than 99% of a wind farm's area is left undisturbed (if somewhat fragmented). In most settings wind farms should not have adverse effects on terrestrial fauna (of course, they are deadly to birds) or on such previous commercial land uses as grazing or cropping. At the same time, assuming that turbines occupy only 1% of land required for the spacing of large turbines would underestimate their overall spatial impact because the noise generated by these large machines requires considerable buffer zones surrounding wind farms. This necessity is no problem in mountains or in regions far from any settlements, but the exclusion of permanent habitation within the buffer zones restricts turbine location in more densely populated European and Asian landscapes.

Many energy projects also include new, deliberately created green belts or buffers designed to provide at least partial visual, noise, and air pollution screens separating them from their surroundings. Such projects are usually site-specific. Some are well planned, others are added as afterthoughts, and yet others are a part of standard construction requirements. For example, India's Ministry of Environment and Forests stipulates that the total green area of the country's numerous new coal-fired power plants is to equal one-third of the total plant land claim, or about 60 ha landscaped and planted for a 1-GW station and almost 120 ha for a large 4-GW plant (CEA 2007).

Finally, many energy production facilities contain land that is entirely undisturbed by their construction; this includes land that was acquired before the project's initiation and that is held in reserve for a possible expansion, land that was deliberately acquired in order to put some dis-tance between a project and the nearest inhabited areas, and, in the case of nuclear power plants, land whose previous uses (unexploited, forestry, cropping, seashore) can continue as long as the land does not include any permanent habitable structures. Some inland water bodies (lakes or reservoirs) associated with thermal power plants also belong to this category: because they are used for water cooling, their area should be counted as a part of the project spatial claim, while most of their former uses are largely unaffected.

As far as the overall impacts of energy infrastructures on land use and land cover are concerned, it should be obvious that they cannot be captured simply by adding up the affected space. Detailed, realistic appraisals require qualitative assessments because energy developments affect spaces ranging from unproductive, barren, hilly surfaces to extensive areas of highly fertile alluvial soils. Similarly, many energy infrastructures have negligible impacts on flora and fauna but others destroy parts of highly biodiverse environments. More often, the ROWs and access roads needed to bring fuels and electricity to distant markets are a major reason for habitat fragmentation, a change that can contribute to loss of biodiversity.

Project Longevities

As for the longevity of specific energy-related land use, it is obvious that only a few structures of modern fossil fuel industries and electricity generation can be called permanent, even if that adjective refers to the span of just

a single century. Mining facilities, oil wellheads, refineries, tanker terminals, fuel storage facilities, and thermal electricity-generating plants are
designed for service spans of 20–40 years, but many of these enterprises
have been around, after upgrading and partial reconstruction, for 50–80
years. What is much more common is that the structures are completely
replaced and modernized but the extraction and generation sites remain.

Many European coalfields were in production (though, for generations,
on a small scale) for more than a century, and some English ones for more
than three centuries, before their operation became uneconomical during
the closing decades of the twentieth century, not because they ran out of
coal (Smil 2010b). California's late nineteenth-century San Joaquin Valley
oil fields are still in production more than a century later: Midway-Sunset
was discovered in 1894, Kern River in 1899 (SJVG 2012). Even more impressive, Baku's oil fields, where modern oil production began in 1846 (a decade
before the United States began drilling in Pennsylvania), are still in relatively vigorous production (Mir-Babayev 2002). And several of the world's
largest oil-producing sites are now more than 60 years old: the Saudi
al-Ghawar field, by far the world's largest oil field, began oil extraction in
1951, and the Kuwaiti al-Burqan site has been producing oil and gas since
1946 (for more details, see chapter 4).

Many coal-fired electricity-generating stations have occupied the same
sites for generations. In 2008 the United States had 10 coal-fired-powered
units that were built during the 1920s, including the Sixth Street Generating Station in Cedar Rapids, Iowa, which began to generate electricity in
1921 and was shut down only in 2010 (SourceWatch 2014). There were an
additional 110 units that began to work before 1950, and plants of similar
age, or even older locations with refurbished generating equipment, can be
found in Europe. And most reservoirs created by dams built during the closing decades of the nineteenth century are still in operation, and most of
those built since World War II will be around for more than 100 years.

Rheinfelden, on the German and Swiss border, was the first large hydroelectric plant in Europe when it was completed in 1898, and after rebuild-
ing between 2006 and 2011 its four new turbines began producing up to
600 GWh/year (Voith 2011). Other well-known large hydroelectric projects
older than 70 years include America's Grand Coulee Dam on the Columbia
River (which entered use in 1942), the Hoover Dam on the Colorado
River (1936), and Ukraine's Dnieper Hydroelectric Station (1932), and the

eventual life span of some of the largest reservoirs will be measured in centuries. That is true even for such impoundments affected by a high rate of silting as Egypt's High Aswan Dam. Negm and co-workers (2010) modeled the reservoir's silting and scouring processes and concluded that the steady value of life span of the dead zone (below the lowest water intake level for the turbines) was 254 years and that of the live zone was 985 years.

On the other hand, many small hydroelectric stations built in China's countryside since the late 1950s as a part of the Maoist quest for inexpensive, mass labor–based solutions to the country's energy shortages silted very rapidly and were abandoned or dismantled after only a few years of unreliable electricity generation (Smil 2004a). Similarly, large numbers of surface or shallow underground mines opened by Chinese peasants (legally and illegally) since the 1950s operated for just a few years before less dangerously produced fuel became available. And the latest extraction technique produces short-lived wells: the average life expectancy of wells drilled for hydraulic fracturing of shales will be no more than 15–20 years (see chapter 4).

Taking these different longevities into account when calculating power densities of specific processes is not easy. The only possible way is to make serial assumptions regarding their ultimate life spans—and such assumptions may have large errors. The redrilling of old reservoirs and the secondary recovery of previously unobtainable oil have extended the lifetimes of many oil fields to three, four, or even five generations, while the initial expectations may have been for just 20–30 years of production. Horizontal drilling and hydraulic fracturing created commercial reserves out of resources that two decades earlier were in the uneconomical category.

In contrast, the availability of cheaper hydrocarbons led to the abandonment of many coalfields (in Europe, Japan, Taiwan) long before they reached their expected life spans. In the Dutch case, that unexpected shift away from domestically produced and imported coal to natural gas from the newly discovered Groningen reservoir took place in just a few years (Smil 2010b). And because we do not have enough accumulated experience with new conversion techniques, assumptions have to be made about the longevity of different forms of solar energy capture and the durability of wind turbines. The standard assumptions are for 20–30 years of service, but the first designs of large commercial wind turbines, introduced during the late 1980s, and even many better models of the 1990s were retired after less than a decade of operation, and the sites were repowered with larger machines.

But these uncertainties do not make any difference for land claims made only by fixed structures. For example, a natural gas liquefaction plant that occupies a relatively small area of coastal (or reclaimed) land will process the gas with the same power density for as long as its annual capacity remains constant. And when calculating the power densities of processes that entail both fixed and incremental land claims, it must be made clear to what specific operations the latter category refers. For example, the fixed land claim of a coal-fired power plant with 1 GW of installed capacity remains constant as long as its operation is not expanded or curtailed. And as long as that plant burns only coal from an adjacent surface mine that taps a seam with reserves good for decades, then its annual incremental (variable) land claims (for coal extraction and the disposal of captured ash and desulfurization slurry) would hardly change and its operating power density, calculated on the basis of annual output, would remain nearly identical.

Land Restoration
Similarly, it is very difficult to offer useful guidance about the probability with which a specific land cover destroyed or altered by energy develop-ment will return (fully or partially) to its predevelopment use. Surface coal mining, one of the most common landscape-altering activities, illustrates the range of possible outcomes. On the one hand, there are some relatively rapid recultivation efforts: land disturbed by surface coal mining may be rather promptly reshaped, re-covered with topsoil, and replanted, and within a decade or two after the seam was extracted the landscape may assume a pleasing aspect composed of a combination of water reservoir (flooding the deepest land cuts) and slopes and flat surfaces replanted with grass, shrubs, or trees, preferably in the more natural, variable, multispecies clumplike fashion rather than (as used to be done) in tree or shrub mono-cultures set out as regular rows of saplings.

An increasingly common way of restoring land and waters claimed by energy projects is by dismantling dams, mostly to improve the upstream access of anadromous fish (Whitelaw and MacMullan 2002). In the United States, the first notable removal dates to 1973 (the Lewiston Dam on the

South Fork Clearwater River in Idaho), and recently accomplished demolitions include that of the 6.6-m-tall Embrey Dam (on the Rappahannock River in Virginia), in 2004, and the 33-m-tall Elwha Dam (on the Elwha River in Washington state), the largest dam removal project so far, which was completed in 2012 (NPS 2013). In Europe the most important recent decision was to remove the Poutès dam on the Allier (a main tributary of the Loire) to allow passage of Atlantic salmon (RiverNet 2012).

In contrast to some notable examples of prompt land use restoration there are many old, derelict energy landscapes (some going back for generations) that have yet to be returned to agriculture, forestry, or recreational use. These areas include old coal-mining heaps, land disturbed by surface mining that was never even leveled or shaped into surfaces suitable for replanting; abandoned oil fields with derelict machinery, burst pipes, and oil pools; and crumbling structures of shut-down refineries and coal-fired power plants. In the United States, there are estimates of up to 500,000 abandoned mines (including all ore and nonmetallic minerals), some 4,000 km^2 of unreclaimed coal mine land, and a data portal showing many details by state (Abandoned Mine Lands Portal 2013).

In between the two extremes of prompt landscape restoration and long-term abandonment is the increasingly common creation of brownfields. After the end of their useful life, many energy sites, much like industrial facilities such as defunct iron and steel mills or chemical and manufacturing (mainly textile and automobile) plants, are simply stripped of all structures and, if need be, the soils are appropriately decontaminated. While some of these newly created brownfields may eventually be replanted, most of them are not renaturalized but turned into another kind of commercial development, most often warehouses, commercial spaces, and apartment blocks.

Concurrent Uses of Space
Space, rather than land, is a more appropriate term here because the two most important examples of concurrent uses do not have any direct footprints on land. PV panels installed on the rooftops of houses and commercial and industrial buildings make no new land claims, and the electricity they supply to the grid (in excess of the needs of the buildings underneath them), by using existing power lines, also helps to reduce additional need for land devoted to transmission from new central generating plants that

would have to be built in the absence of new rooftop PV capacities. And by far the most important example of concurrent space uses by the traditionally most important renewable energy conversion does not refer to land but to reservoirs impounded by large dams.

In calculating the power densities of hydroelectric generation, reservoirs' entire areas are used as a denominator, but they also supply drinking and industrial water for cities and irrigation for farming, provide flood control (whose economic benefits, namely, preventing or reducing major downstream inundations, may be greater than the value of the electricity generated), improve upstream navigation, and offer new opportunities for recreation as well as for freshwater aquaculture, one of the fastest-expanding modes of modern food production. On the other hand, the negative environmental impacts associated with large reservoirs may actually extend their spatial impact.

When new reservoirs flood arable land and established settlements and industries, new areas must be reclaimed to maintain food production and to accommodate resettled populations, while irrigation may result in excessive salinization and abandonment of the affected land. These additions and losses are directly or indirectly attributable to the creation of large reservoirs, but we have no satisfactory way to weigh them against the just noted multipurpose uses and to come up with a net adjustment for a specific reservoir area (which in some climates can show significant inter- and intra-annual fluctuations).

Uses that have no effect on the original design of a reservoir can be ignored: swimming, boating, or a limited diversion of drinking water (amounting to a fraction of natural evaporation from a reservoir) would be in that category. But if the volume of the reservoir was deliberately designed to store water for both electricity generation and seasonal irrigation, or if a reservoir's exclusion area has been made larger to accommodate periodic floods, then it would be logical to apply appropriate correction factors and not to use a reservoir's entire area as a denominator in calculating power densities.

Applying spatial discounts for other uses would raise the average power densities of electricity generation, but such a correction would be relatively easy to do (although still arguable) only in the case of clear and measurable dual use, for example when data on water withdrawals for irrigation could be compared with the water volume used for electricity generation. In the case of multiple and nonconsumptive uses (recreation, aquaculture), it would be arbitrary. Moreover, even if some acceptable generalized way to correct for non-electricity-generating uses could be found, the power density's order of magnitude would not change; it would undergo merely a fractional adjustment, and hence it might be argued that it is sensible to ignore multiple reservoir uses.

Multiple uses may also mean that the loss of the land claimed by a reservoir may be more than compensated for by the creation of new productive land. Egypt's High Aswan Dam is a perfect illustration of this reality. The dam created a large lake that flooded mostly desert wasteland in southern Egypt and northern Sudan and caused the loss of only a narrow strip of previously cultivated land along the river that had benefited from silt deposition by annual flooding. At the same time, irrigation water from the reservoir enabled Egypt to expand its cultivated area in the Delta and along the Nile Valley and to convert many previously single-cropped fields to double- and triple-cropping: the overall gain of farmland amounted to about 50%, and more food production was added by the reservoir's commercial fishery (Abu-Zeid and el-Shibini 1997). On the other hand, silt deposition in the reservoir cut off the downstream farmland from its annual addition of nitrogen-rich sediments, a loss that has to be compensated for by applications of synthetic nitrogenous fertilizers.

Perhaps the most commonly cited example of concurrent land use that involves new renewable energy conversions (and that is often stressed by their proponents) is the case of wind-generated electricity: concrete turbine foundations occupy only small areas (on the order of 15–25 m^2) and, except for permanent access roads, the surrounding landscape can be used for grazing or crop cultivation. Moreover, depending on the crops grown and wind-site royalties, a farmer may earn more from the latter than from the harvest of the former. As Lovins (2011b, 2) put it, "saying that wind turbines 'use' the land between them is like saying that the lampposts in a parking lot have the same area as the parking lot: in fact, ~99% of its area remains available to drive, park, and walk in."

That is a correct statement—but it is also a complete misunderstanding of the power density concept. Power density informs us about the energy flux that can be usefully derived from a given area by a particular conversion, regardless of how close or how far apart the individual converting or extracting facilities may be, and regardless of the commercial activity (if any) taking place in most of the area that is not occupied by structures or infrastructures indispensable for extraction or conversion. Consequently, using a wind farm's entire area in the power density's denominator is not conceptually different from using the gross land claims of other energy-producing activities, even though the actual disruptive footprint of their infrastructures adds up to only a very small share of the overall claim. For example, in terms of the ratio of occupied to unaffected land, it is very similar to using the entire area of an oil field, where only a fraction of land is taken up by well pads and access roads and most of it is planted to crops or is grazed by cows: many American landscapes, from California to Texas, offer numerous examples of this common reality.

Growing crops or grazing cows has nothing to do with the fact that even in not very sunny mid-latitudes, PV cells will always produce electricity with a higher power density than wind turbines, or that even old oil fields dotted with pumps will have a higher power density of fuel extraction than converting soybean harvests to biodiesel. And the power densities of roof-top PV panels obviously do not tell us how much additional land they occupy (none at all) but what the existing or potential electricity-generating capability of such installations is, particularly when those rates are compared to average electricity demand rates of buildings underneath the panels.

Boundaries and Technical Advances

Two critical considerations apply to all power densities: the boundary problem in calculating the rates, and technical advances that extend the realm of possible achievements. The first reality creates numerous uncertainties: because there are no binding rules for the inclusion or exclusion of analyzed components, different choices of system boundaries will result in substantially different outcomes. Again, this is not a challenge unique to power densities for it is repeatedly encountered in analyses of energy costs (embedded energies). They can be narrowly limited to a specific process (smelting steel) or they can include the energy costs of as many preceding steps as is practicable (mining and transporting iron ore, producing coke and natural gas, mining limestone, producing temperature-resistant furnace linings, etc.).

The resulting differences in final energy costs are commonly two- or threefold. For example, the energy costs based on an input-output analysis of the steel industry were twice as high as those calculated by basic process analysis (Lenzen and Dey 2000), and similar comparisons showed the energy cost of structural iron and plywood used in Swedish apartment buildings to be three times as high when input-output data were used (Lenzen and Treloar 2003). Some calculations presented in chapters 3–5 will show that differently set analytical boundaries can result in even greater disparities for power densities, some easily as large as an order of magnitude.

This is particularly true when areas of requisite distribution (transportation and transmission) infrastructures of the fossil fuel industries and of all forms of electricity generation are added to spaces claimed by extraction and power plant facilities. Surfaces that are actually transformed by high-voltage transmission lines, the most extensive of these distribution infrastructures, are minimal: the foundations of transmission towers occupy only a small fraction of the ROWs set aside for the lines, and the land underneath the lines can be used as before if it was grazed or cropped, or it can be converted to plant nurseries or Christmas tree plantations. Similarly, land above buried pipelines can be used for grazing or can be planted to seasonal crops; in Canada, although not in the United States, even trees are allowed in a pipeline's ROW as long as they are less than 1.8 m tall and are at least 1 m away from the line, and the only land whose use has been grossly altered is for pumping and compressor stations placed along the line.

Of course, as already noted, similar situation arises with wind turbines, and there is no accepted norm to deal with this large disparity between transformative (footprint) and spacing (ROW) land requirements. Counting only the former clearly underestimates the overall impact because land use within ROW corridors (or within spaces occupied by large wind farms) is obviously restricted, and using the overall spacing or ROW power density conveys the fundamental limits on the capacity of machines, lines, or pipes that can be accommodated within a unit of land. At the same time, an ROW claim is not obviously equivalent to such (if only temporarily) destructive transformations as surface coal mining or the impervious surfaces of many energy facilities.

Perhaps the best way to deal with the inescapable boundary problem is to follow these three basic rules: to make clear what goes into a particular power density calculation; where appropriate, to present alternative values of power densities within narrower and wider (and always properly explained) boundaries; and to make sure that comparisons of different modes of energy extraction, conversion, and use are done for accounts that have been prepared in identical, or at least very similar, ways. I will try to follow these precepts in this book. Finally, a few comments about the evolving value of the power densities of energy production and use are in order.

The power densities of renewable energy flows and energy densities of fossil fuel resources are fixed. The first category is circumscribed by the amount of solar radiation reaching the Earth and its subsequent transformation into wind (via differential heating of the planet's surfaces), flowing water (via evaporation), and phytomass (via photosynthesis). Stores of fossil fuels present a fixed outcome of long underground transformations involving elevated temperatures and pressures (in theory, these resources are not finite, but their rate of formation is negligible compared to our current rates of their extraction). In contrast, the power densities with which these resources are converted to useful energies are—within the natural and thermodynamic constraints—anthropogenic, and as such they evolve, driven by a combination of technical and managerial advances.

The conversion of phytomass to heat and light began hundreds of thousands year ago (dates for the earliest controlled use of fire remain uncertain), the conversion of running water and wind (by waterwheels and windmills) dates, respectively, to more than two millennia and about 1,200 years back (Smil 1984). All of these very low power densities and rates began to rise only with the advent of modern high-yield cropping and forestry and with the invention and commercialization of water turbines (starting in 1830s) and modern wind turbines (in the 1980s). The direct conversion of solar radiation was impossible until the first deployment of PV cells on Earth's satellites in the 1960s, and the power densities of solar electricity generation have been rising with improving cell efficiencies.

Similarly, the power densities of fossil fuel extraction and thermal electricity generation have increased as a result of better mining, drilling, and recovery techniques and the rising efficiencies of boilers, nuclear reactors, steam turbogenerators, and gas turbines. Moreover, some of these innovations (surface coal mining, deep and horizontal drilling, and hydraulic fracturing) made it possible to recover resources that were previously considered uneconomical, including through the surface mining of thick but low-quality coal deposits more than 300 m belowground, the recovery of hydrocarbons in reservoirs deeper than 3 km, and accessing nonconventional resources of oil and natural gas bound in shales.

This typological detour and the brief notes on assorted caveats were necessary to appreciate that both of the variables used to calculate power densities have a wide range of qualitative attributes. This reality has many counterparts in endeavors aimed at analyzing and understanding other complex phenomena, be they (as I noted when pointing out a quality-free common denominator of food energy densities) studies of human nutrition and average per capita dietary intakes or (to cite very different examples) efforts to understand educational achievements and economic inequality.

There are, obviously, enormous qualitative differences among PhD degrees from different universities and in different disciplines, or among sources of income measured in the same currency but coming from such qualitatively disparate sources as a comfortable trust fund or three poorly paying part-time jobs. These realities complicate, but never disqualify, the measures we have to use to derive an all-encompassing understanding of complex phenomena. The power density rate is a powerful explanatory variable, but it should be assessed in conjunction with other factors that codetermine the capacities and performances of a specific resource extraction or conversion method or the overall system capability.

In environmental terms, power density is about claiming space: land use intensity (m^2/W) is its obvious inverse. But there are other intensities to consider, above all the intensity of water use (g H_2O/J) and carbon intensity (g C/J), a marker of the human interference in the global biogeochemical carbon cycle that quantifies the emissions of CO_2, the dominant anthropogenic greenhouse gas. And even before considering water or carbon, costs might come first to mind. The power density and specific energy production costs of fossil fuels have an obviously significant inverse correlation, but there is no simple general causality.

Hydroelectric generation, a ubiquitous example of a low-power-density conversion, produces some of the world's cheapest electricity, while liquid metal fast breeders would produce electricity with a much higher power density than rooftop PV cells, but all countries that were initially engaged in operating experimental breeder reactors (the United States, the USSR, and then Russia, France, Japan, and India) found their costs so prohibitive that they stopped (the United States and UK by 1994, France by

2009), curtailed, or suspended further development aimed at commercial application (Cochran et al. 2010). But these two examples are uncharacteristically clear-cut. In most other cases cost considerations are subject to some profound uncertainties.

All kinds of modern energy conversions have benefited from often generous, and longlasting, subsidies (Gerasimchuk et al. 2012; Laderchi, Olivier, and Trimble 2013; OECD 2013), and a proper accounting of these hidden costs could easily swing around many comparisons of specific energy modes. For example, the most comprehensive comparison of the levelized cost of electricity generation shows that in 2013, US federal tax subsidies could cut the cost of crystalline rooftop PV arrays to as low as $117/MWh (the unsubsidized cost was as high as $204/MWh), making it in many places competitive with coal-fired generation, whose unsubsidized cost ranged from $65 to $146 (Lazard 2013).

Similarly, a comparison can be turned around by including only one or two ignored externalities: the adoption of fairly high carbon taxes designed to account for the long-range effect of global warming would make coal-fired plants—now about $100/MWh, and quite competitive with most forms of renewable electricity generation—instantly uncompetitive. Such shifts would require different margins in different countries, and that is why I will not offer any comparative lists claiming that wind-generated electricity is always cheaper than any thermal generation or that PV is the cheapest alternative. Instead, I will point out a number of technical and environmental realities that should be considered along with power densities.

3 Renewable Energy Flows

This chapter offers a systematic stroll through renewable energies that begins with solar radiation and its three great derivatives—winds, flowing water, and phytomass produced by photosynthesis—and concludes with a brief assessment of geothermal energy, the Earth's most important nonsolar energy flux.

When classified by their origin, renewable energy flows belong to just three categories. By far the most important is solar radiation (direct and diffuse) and its diverse transformations, manifested as wind (generated by pressure differences resulting from the differential heating of the Earth's surfaces), a vertical temperature gradient in the ocean (created by the heating of the surface layer above the isothermal deep water), ocean currents (resulting, again, from differential heating of the liquid mass), the kinetic energy of streams (created by the Sun-driven water cycle of evaporation, precipitation, and runoff), and plant mass (phytomass, created through photosynthesis).

The second category of renewable energy flows is energized by the decay of radioactive elements in the Earth's crust and by the flow of basal heat from the planet's hot interior, and manifests as an omnipresent but low temperature gradient and, particularly along tectonic plate boundaries, as high-temperature-gradient flows of hot water and steam that can be used for space heating or electricity generation. The last category of renewable flows also has only a single entry, as the gravitational interplay of the Earth and its orbiting satellite creates tides that are highly predictable but that reach impressively high differentials only in a few regions.

I will concentrate on major land-based fluxes whose conversions have been indispensable in the past, that are increasingly exploited today, and that have the greatest practical promise for future innovations. These flows include the conversion of solar radiation into heat (the most common application is to heat water) and electricity (done mainly by photovoltaic [PV] cells or by concentrating solar radiation to heat a working medium in a standard thermal power plant); phytomass (used for millennia as the dominant source of heat and light in the form of fuelwood, charcoal, and lamp oils, and more recently also converted to liquid fuels, above all to ethanol and biodiesel); hydro energy (a leading source of mechanical energy during the preindustrial era, produced by waterwheels, and now a major global supplier of electricity generated by water turbines); and wind (converted in the past into mechanical energy by windmills, and today into electricity by wind turbines).

Despite its large aggregate global potential, geothermal power must be classed in a less consequential category, and although the commercial exploitation of ocean currents, ocean thermal differences, and tidal energy has some tireless proponents, these energy sources will not be considered here: they are the least promising renewable energy alternatives, their power densities (per unit of ocean surface) are inherently low, and their land claims (for associated infrastructures) would be minimal. My systematic review will clearly establish a hierarchy of renewable energies based on their average, as well as peak, power densities. These findings are used later in the book when I contrast the typical power densities of modern ways of energy production with the prevailing, and changing, power densities of energy use in high-energy urban societies.

Solar Radiation and Its Conversions to Heat and Electricity

The Sun is situated in a spiral arm of our galaxy, with its closest neighboring star, α Centauri, 4.3 light-years away. Our star is a dwarf (size class V) belonging to a rather common spectral group (G2), whose radius (696.7 Mm) is more than 100 times that of the Earth, and it keeps the planet orbiting at a mean distance of 149.6 Gm (Mullan 2010). Hydrogen makes up about 91% of the Sun's huge mass, and its energy is produced in its core mostly by the fusion of protons, forming helium, which consumes about 4.4 t of its mass every second. The Stefan-Boltzmann equation states that the radiant flux (F) must be proportional to the fourth power of temperature (box 3.1).

Box 3.1

Radiant flux

$F = \sigma T^4$

σ (Stefan-Boltzmann constant) = 5.67×10^{-8} W/m²/K⁴

$T = 5,779$ K

$F = 5.67 \times 10^{-8} \times (5,779)^4$

   = 63.24 MW/m²

This means that the Sun's isotropic radiation is 63.2 MW for every square meter of its photosphere.

This granulated layer radiates along a wide spectrum whose wavelengths range from less than 0.1 nm (γ rays) to more than 1 m (infrared radiation). The peak radiation—dictated by Wien's displacement law, $\lambda_{max} = 0.002898/T$ (K)—is at about 500 nm, close to the lower limit of green light (491 nm), and about two-fifths of all energy is radiated in visible wavelengths between 400 nm (deep violet) and 700 nm (dark red). Ultraviolet radiation carries about 8% of all solar energy, infrared the rest (53%). Visible wavelengths energize photosynthesis (it proceeds mainly by the means of blue and red light, with green reflected) and are sensed by organisms ranging from bacteria to humans: our vision is most sensitive to green (491–575 nm) and yellow (576–585 nm) light, with the maximum visibility at 556 nm. But the heating of the biosphere is done mostly by infrared radiation at wavelengths shorter than 2 μm.

Once the radiation leaves the Sun's photosphere it travels virtually unimpeded through space, and its power density at the top of the Earth's atmosphere is easily calculated by dividing the star's total energy flux (3.845 $\times 10^{26}$ W) by the area of the sphere whose radius is equal to the planet's mean orbital distance of 149.6 Gm. This rate—1,367 W/m²—is known as the solar constant, although the rate has both short-term and longer-term deviations (de Toma et al. 2004; Foukal 2006). The mean value of satellite observations is 1,366 W/m², with brief declines of 0.2–0.3 W/m² (caused by the passage of large sunspots across the Sun) and periodic undulations (caused by the 11-year solar cycle), with the peaks as high as 1366.9 W/m².

Although the Earth is a rotational ellipsoid rather than a perfect sphere, dividing the solar constant by four (the difference between the area of a circle and of a sphere of the same radius) yields a fairly accurate mean of the solar radiation (341.5 W/m^2) that would reach the rotating planet if it were a perfect absorber (black body) and had no atmosphere. In reality, the Earth's atmosphere absorbs about 16%, and hence even without any clouds the annual mean irradiance would be no more than 287 W/m^2. Clouds absorb another 3%, reducing the mean to about 278 W/m^2. But the atmosphere is also a reflector of incoming radiation; long-term satellite measurements have confirmed that the Earth's albedo (the share of the reflected radiation) averages almost exactly 30%, with obvious seasonal fluctuations between the northern and southern hemispheres.

The Earth's atmosphere reflects 6% of the incoming radiation, clouds reflect 20%, and continental surfaces and water account for the remaining 4%. This means that about 55% of incoming radiation could be absorbed by an average square meter of a perfectly nonreflecting horizontal ground surface, the rate that translates to about 188 W/m^2. Actual irradiance is subject to daily and seasonal variations that can be perfectly predicted for any site on the Earth as the planet rotates on a tilted axis; it is also subject to the much less predictable mesoscale influences of cloud-bearing cyclonic systems, and to highly unpredictable local cloudiness. The values for the total annual global irradiance are, obviously, a function of latitude and cloudiness. Below I cite just a few representative rates calculated from daily measurements (Ineichen 2011; NASA 2013).

Annual Irradiance

The rate for Berlin, representative of populated northern hemisphere mid-latitudes influenced by regular cyclonic flows, is just 116 W/m^2 (an annual energy total of 1.016 MWh/m^2); the rate for Dublin, within the same zone, is only about 100 W/m^2. London is very close to Berlin (110 W/m^2), while Paris gets around 125 W/m^2. Murcia, in Spain, representative of sunny Mediterranean locations, receives 196 W/m^2, similar to Athens and a bit more than Rome, at about 175 W/m^2. The southern states of America's Great Plains experience the same range of average irradiance (Tulsa in Oklahoma receives about 180 W/m^2, San Antonio in Texas about 200 W/m^2).

At 266 W/m^2 (2.328 MWh/m^2 in a year), Tamanrasset in southern Algeria indicates the maxima receivable in the planet's sunniest climates of the arid subtropical belt. The only larger city with a slightly higher long-term rate is Nouakchott, the capital of Mauritania, which receives 273 W/m², while the Saudi capital, Riyadh, receives slightly less, about 251 W/m². The largest relatively heavily populated regions with average rates above 200 W/m² are in the US Southwest (with Los Angeles and Phoenix receiving 225 W/m²) and Egypt's Delta (Cairo at 237 W/m²). In contrast, frequent cloudiness keeps the rate for such tropical megacities as Singapore and Bangkok well below 200 W/m².

Differences in monthly insolation averages depend on latitude and cloudiness: in Oslo the difference between January and June is 16-fold; in Riyadh it is only twofold. These differences are reflected by actual PV electricity generation: in 2012 the German output was 4 TWh in May and just 0.35 TWh in January, an order of magnitude disparity (BSW Solar 2013). The annual variability of irradiance is significantly larger in cloudy climates and in mountainous regions: fluctuations of total irradiance are as high as 7% in Berlin and Zurich, 2% in Murcia and Tamanrasset. Naturally, monthly differences can be much larger. For example, in January 2013 the Czech Republic averaged 50% less sunlight than in January 2012 (Novinky 2013). During cloud-free days the highest daily global irradiance is a perfectly predictable function of a calendar day and latitude; naturally, daily minima in the northern hemisphere occur during the winter solstice and maxima occur six months later. Theoretical expectations of noontime maxima are as high as 1,065 W/m² (the solar constant minus atmospheric absorption and reflection).

Indeed, the highest recorded maxima in subtropical cloud-free deserts come within a small fraction of 1% of that value. For example, the noontime May and June hourly maxima from Saudi Arabia are 1,059 W/m² for Dhahran on the Gulf coast and 1,056 W/m² for Riyadh in the interior (Stewart, Dudel, and Levitt 1993). Irradiance delimits the range of power densities that can be harnessed as heat or converted by PV cells to electricity. There is an order of magnitude difference between annual averages of less than 100 W/m² in cloudy temperate mid-latitudes and the just noted maxima above 1,000 W/m² available for one to three hours a day during the sunniest spells in the great subtropical desert belt that extends from the Atlantic coast of Mauritania to North China and that has its (much more circumscribed) western hemisphere counterparts in the US Southwest and northwestern Mexico, and in the southern hemisphere it prevails throughout large parts of Australia and in a narrow strip along the coast of South America.

Conversions of Solar Radiation

Life's evolution and the biosphere's dynamics are determined by the levels and variations of irradiance. Irradiance energizes photosynthesis and warms the atmosphere, waters, rocks, and soils (creating pressure differences and hence powering the global air circulation, and evaporating moisture and hence powering the global water cycle), as well as bodies of organisms (critical to keep optimal temperatures for enzymatic reactions) and surfaces of buildings (every aboveground structure that shelters is solar heated). Direct solar radiation has also been used for millennia to evaporate salt along seashores, to dry crops, to sunbake clay bricks, and to dry clothes. In cold climates the best design tries to maximize solar gain in passive solar buildings—and, taking some ancient lessons, to minimize it in hot environments (Athienitis 2002; Mehani and Settou 2012).

Not long after modern indoor plumbing became available, solar radiation began to be used to heat water in simple rooftop collectors. This practice has become increasingly more efficient and is now quite common in many sunny countries (Mauthner and Weiss 2013). Improvements in plant productivity aside, the most important active step in harnessing irradiance has been the development of solar electricity generation, primarily based on the PV effect but also using concentrated solar power to run steam turbines. These conversions have power densities higher than those of any other means of harnessing renewable energy flows; moreover, PV efficiencies have been gradually improving, and further significant gains are certain.

The power densities of solar conversions have several unique attributes: they belong to two distinct categories; calculations for nearly all of them are done for tilted rather than for horizontal surfaces; and some of them refer to active (adjustable) areas rather than to fixed level surfaces. Most tilted panels on roofs on large solar farms are fixed in one position: optimal angles from horizontal are easily calculated (Boxwell 2012). The best full-year angle for my latitude (Winnipeg is 50°N, the same as London or Prague) is 41.1 degrees, and that position will capture about 70% of radiation compared to a tracker; for Tokyo (35°N) it is 29.7 degrees; and, obviously, fixed panels should remain horizontal at the equator. Panels are usually adjusted according to season (twice or four times a year), while full tracking has so far been reserved for some commercial installations.

The two distinct categories are land-based systems (large-scale PV plants and concentrated solar power harnessing–irradiance with a field of helio-stats) and rooftop installations (both for solar heating and for PV electricity generation). In the future a third category might become important as ver-tical surfaces—mostly suitably oriented building walls—can be either clad in PV panels or glazed with PV glass unit windows (Pythagoras Solar 2014). With land-based systems, solar power densities are both similar to and dif-ferent from those of wind-powered electricity generation. PV arrays, much like wind turbines and their associated access roads and structure, will not completely cover the land claimed by the project, but their degree of cover-age will be much greater, and in most cases the PV arrays will be fenced.

In contrast, the power densities of rooftop installations (whether they are for heat or for electricity) are in a special category as no other energy converters are routinely placed on tops of buildings and they do not claim any new land surface. The same is true of wall-based PV or PV-integrated windows. Moreover, during the periods when rooftop PV modules generate more electricity than can be used by the buildings on whose roofs they are situated, they can send the surplus to the grid. Rooftop thermal and PV solar conversions are thus doubly land-sparing, and as they already have power densities higher than that achievable with the harnessing of any other renewable energy flux, they deserve to be widely promoted and adopted, although the intermittency of the flux continues to pose non-trivial challenges to any system that would raise their contribution to a rela-tively high level.

Solar Heating

Water heating using rooftop collectors is highly efficient and affordable. With highly selective coatings, the latest designs of flat-plate collectors (typical panels circulating water cover about 2.5 m^2 and are less than 8 cm thick) have an absorptivity of 95% (Bosch Thermotechnology 2014; Stiebel Eltron 2014). Evacuated glass tubes have a similarly high absorptivity. They have vacuum surrounding a heat pipe fused to an absorber plate, and they are more efficient, particularly in colder climates and in winter in sunny climates (Silicon Solar 2008). Their efficiencies (when compared for insola-tion at 1,000 W/m^2) differ with the desired temperature: for pools (a water

Figure 3.1
Rooftop water heating using evacuated tubes. Photo available at Wikimedia.

temperature up to 25°C) they have efficiencies of 60%–80%, whereas for domestic water heating (water at 45–60°C), the efficiency figure falls to 40%–75% (Thermomax Industries 2010). Australian studies show that evacuated tubes are about 50% more efficient than flat plates in summer and 80%–130% more efficient in winter (Hills Solar 2008).

During the midday hours in such sunny climates as California or Mediterranean Europe, the high absorptivities of flat plates and evacuated tubes translate into power densities of heat collection in excess of 900 W_t/m^2. During those times water heating may proceed with power densities higher than 700 W_t/m^2, at rates that are unequaled by any other commercial renewable energy conversion (fig. 3.1). The annual power densities of water heating in sunny climates are much lower but, again, much higher than those of any commercial renewable energy conversion: they can be as high as 110 W_t/m^2 in Israel or Arizona, while in temperate climates they will generally be no higher than 40–50 W/m^2. For example, detailed German data show that by the end of 2012, Germany had 16.5 Mm^2 of solar heating surfaces whose output was about 685 MW_t (BSW 2013), and that implies an average power density of 41.5 W_t/m^2.

Obviously, larger storages extend hot water availability but come at a higher cost. Roof placement is usually available, but shading (total or partial) by nearby trees, buildings, and structures cannot always be avoided. Combined systems can capture irradiance for both space and water heating. The worldwide total of solar heat collectors (dominated by small rooftop heaters) is in the tens of millions. Mauthner and Weiss (2013) estimated that by the end of 2012, solar thermal collectors had an aggregate area of 383 Mm² (compared to 125 Mm² in 2005), a total capacity of 268 GW$_t$, and an annual output of 225 TWh; that implies a globally averaged power density of 67 W$_t$/m².

Power densities between 40 and 100 W/m² are superior to the power densities of all household-based renewable energy conversions. That is why properly scaled distributed units are a perfect choice for household water heating: in sunny climates they can cover moderate daily needs even without any voluminous hot water storage, and in less sunny climates they can make a significant contribution. Modern water heating systems do not have any extraordinary material requirements, are widely affordable, and can operate reliably for a long time. A rational, systemwide approach to energy use would promote the widespread adoption of distributed rooftop solar water heaters for households and smaller-sized commercial and industrial buildings in any suitable climate. But the promotion of these high-power-density distributed converters has become overshadowed by the rapid (and in some instances questionable) growth of PV-based electricity generation.

Photovoltaics

The conversion of solar radiation to electricity was discovered by Edmund Becquerel in 1839. The first experimental PV cells were made in 1877, but commercialization of the process began only in 1954 with the production of the first silicon solar cells at Bell Laboratories. In 1962 Telstar, the first commercial telecommunications satellite, opened the way to PV-powered space vehicles, but terrestrial applications took off only during the 1990s (Smil 2006). They have been driven by an increasing interest in carbon-free electricity generation, but the rapid diffusion of PV systems began only with the adoption of costly subsidies (in the form of guaranteed long-term feed-in tariffs) in some countries.

PV modules and arrays are now mass-produced both for household rooftop installations and for large-scale industrial projects, with modules designed for nominal irradiance, at either 1,000 W/m² or 800 W/m², and at an ambient temperature of 20°C or 25°C. Increasing irradiance raises their current and power output but has a much smaller effect on the voltage; increasing the cell temperature brings a significant decline in voltage and also reduces cell output, efficiency, and expected duration. The deployment of PV cells with a conversion efficiency of at least 10% would produce peak power densities of PV modules in the range of 80–100 W_p/m^2 during a few midday hours; with a 15% conversion efficiency the rate would rise to 120–150 W_p/m^2. Consequently, approximate calculations of aggregate PV performance should not assume less than 100 W_p/m^2 for newly installed arrays.

Average annual power densities of PV are calculated by multiplying measured irradiance by the mean efficiency of modular cells adjusted for their specific performance ratio (the difference between their actual and theoretical output). As already noted, the first number is essentially fixed for a given location (with a relatively small interannual variation), while the other two parameters have been steadily increasing through innovation. In 2014 the best research cell efficiencies were as follows: emerging techniques (organic, perovskite, and dye-sensitized cells), 8.6%–17.9%; thin films, 13.4%–23.3%; crystalline silicon cells, 20.4%–27.6%; and multijunction cells, 26.4%–44.7% (NREL 2014). The actual field efficiencies of recently deployed PV cells have been much lower. The conversion efficiency of single crystalline modules, the oldest but still the most efficient PV conversion technique, averages 10%–12%; that of cheaper polycrystalline silicon cells is now close behind, at 10%–11%; string ribbon polycrystalline silicon delivers 7%–8%; and amorphous silicon (vaporized silicon deposited on glass or stainless steel) will convert no more than 5%–7% of irradiance into electricity.

Assuming, again, an average 10% efficiency would result in a fairly representative range of average power densities for the plants operating at the beginning of the second decade of the twenty-first century. Those densities would range from less than 10 W_e/m^2 in cloudy mid-latitudes (Atlantic Europe, the Pacific Northwest) to more than 15 W_e/m^2 in sunnier climates, and would peak at around 25 W_e/m^2 in cloud-free subtropical deserts. These rates are applicable for small modules installed on roofs on the ground; the power densities are lower for large ground-based installations because additional land is required between tilted PV arrays to avoid shading and to provide access for servicing the modules, for roads, for inverter and transformation facilities needed to access the grid, and for service and storage buildings.

Installations with tracking assemblies require even more additional land per PV module to avoid shading. Consequently, anywhere between 25% and 75% of a solar park area will actually be covered by the modules, while for the PV field alone the cell assemblies typically cover 75%–80% of land. But it should be noted that many large solar parks often acquire or lease much larger areas intended to accommodate possible future expansion, and those areas should not be counted when calculating the power densities of actually operating projects. Other projects claim more land outside their PV array fields to create environmental buffers or to provide corridors for wildlife. The inclusion of those areas in the land denominator is, as with many similar land claims made by other energy installations, arguable.

The general procedure for calculating the power densities of ground-based PV installations is thus quite straightforward: average irradiation (I) is converted from Wh to W, and the result is multiplied by the conversion efficiency (η) and an appropriate performance factor (p). For the average US insolation of 1,800 kWh/m², a cell efficiency of 10%, a performance factor of 0.85, and 50% of the ground covered by modules, the result is almost 9 W_e/m², while for a cell efficiency of 15% this value would rise to 13.1 W_e/m²:

Power Densities of Large Ground-Based Projects

The most important correction in calculating actual solar power densities is the adjustment for their relatively low capacity factors. The capacities of new solar projects are always listed in terms of rated peak power (MW_p), the performance achievable only by perfect conversion during the time of highest irradiance. As expected, average capacity factors correlate with total irradiance: in places where it is less than 150 W/m² they will be below 12%,

Box 3.2
Power density of a PV module

$I = 1,800$ kWh/m²

1.8 MWh/m² × 3,600 = 6.48 GJ/31.5 Ms = 205.5 W/m²

205.5 W/m² × 0.10 × 0.85 = 17.5 W/m² × 0.5 = 8.7 We/m²

for the insolation between 150 and 200 W/m^2 they will range up to 20%, and in the sunniest locations with irradiance in excess of 200 W/m^2 they will be up to 25%. Actual performance data show that even in sunny Spain, most plants have capacity factors of less than 20%, and in cloudy temperate climates that indicator will dip below 10%. In addition, only about 85% of a PV panel's DC rating will be transmitted to the grid as AC power: these performance ratios vary, but in the best systems they should always be above 80% and should approach 90%. Several notable examples of large PV plants illustrate actual power densities.

In 2008 Olmedilla de Alarcón (Cuenca, Castile–La Mancha) became temporarily the world's largest solar park, with an installed capacity of 60 MW_p. Olmedilla's total area of 283 ha of fixed panels and an annual generation of 85 GWh (or an average power of 9.7 MW) translate to a power density of about 3.4 W_e/m^2 and an average capacity factor of just 16%. Another large plant completed in 2008, the Portuguese Moura (46 MWp of installed capacity, 88 GWh or 10 MW of average power) has a capacity factor of nearly 22%. The plant has both fixed and single-axis tracking panels (covering a total of 130 ha), and its power density is 7.7 W_e/m^2. When it was finished in 2011, Sarnia (Ontario) was the world's largest PV plant; its installed capacity is 97 MW_p and its annual generation of 120 GWh comes from 1.3 million panels covering 96.6 ha, while the plant's entire area claims 445.2 ha (Clean Energy 2013c). This prorates to average annual power densities of 14.2 W_e/m^2 for the modules and 3 W_e/m^2 for the total land claim. With an irradiance of about 180 W/m^2, this puts Sarnia's average conversion efficiency at less than 8% and its capacity factor at 14%.

Another large project completed in 2011, Germany's Waldpolenz (about 20 km east of Leipzig, on the site of a former Soviet East German air base) has a peak capacity of 52 MW, an annual generation of 52 GWh (5.94 MW), a total panel area of about 110 ha, and a total site of 220 ha (Juwi Solar 2008). These specifications yield a power density of 5.4 W/m^2 for the module field, 2.7 W_e/m^2 for the entire plant area, and an annual capacity factor of just 11.4%. Perovo (in western Crimea) has a peak power of 100 MW, generates 132.5 GWh (averaging 15.1 MW) from 200 ha of panels (Clean Energy 2013b), and has a power density of about 7.6 W_e/m^2 and a capacity factor of 15%. Agua Caliente, in 2013 the largest project in North America, on 960 ha in Arizona, has one of the world's highest average annual irradiation rates (2.45 MWh/m^2), a capacity of 290 MW_p, and

Figure 3.2
California Valley solar ranch. © Proehl Studios/Corbis.

generates 626.2 GWh/year (Clean Energy 2013a). This yields a high capacity factor of 24.6% and a power density of 7.45 W/m². The California Valley Solar Ranch (250 MW$_p$, 482 GWh, load factor of 22%) occupies 60 ha and has an operating power density of 9.2 W$_e$/m² (CVSR 2014; fig. 3.2).

Consequently, the largest PV projects now operate with power densities of roughly 3–9 W$_e$/m². Because smaller projects use similar or identical PV cells, it is not at all surprising that their power density range is pretty much the same. McKay's (2013) listing of such projects in Italy (with installed capacities between 1 and 10 MW and a capacity factor about 16%) shows the range of 4–9 W$_e$/m²; for projects in Spain (projects rated at 7–23 MW, with load factors between 16% and 23%) the power density range is 4–11 W$_e$/m²; for the UK (projects averaging about 5 MW, with an average load factor of about 11%) it is just between 4 and 5 W$_e$/m²; and for the ground-based US installations the range is from just 3.8 W$_e$/m² for a two-axis 2.1-MW tracker in Vermont to 11.43 W$_e$/m² for a fixed 250-kW installation in Florida. To minimize capital costs, most of the large PV projects use relatively inefficient, less costly, thin-film Cd-Te cells. In contrast, the highest capacity factors would be around 30% with double-axis tracking in the sunniest locations in the US Southwest (Madaeni, Sioshansi, and Denholm 2012).

Rooftop and Facade PVs

In Germany, the world's leader in harnessing solar radiation for electricity, most PV cells are not massed in large solar parks but rather on rooftops, installed by homeowners and businesses in response to feed-in tariffs guaranteeing high electricity prices for 20 years. In 2011 ground-based PV cell arrays in large solar parks accounted for only 28% of Germany's installed capacity; its largest share, 38%, was in medium-sized installations (10–100 kW_p) on the roofs of multifamily dwellings, schools, offices, farms, and small businesses; 23% of all PV modules were larger (more than 100 kW_p) units on the roofs of industrial enterprises, and 10% were on the rooftops of private residences (Wirth 2013).

These rooftop PV modules come in sizes from roughly 0.5 m^2 to 3 m^2. Most modules have 36 cells in series, with a maximum power voltage of 15 V and a maximum power current of 3 A. Changing irradiance affects their current and power output, but the voltage varies very little, facilitating battery charging. In 2012 Germany had nearly 1.1 million solar installations serving 5.2 million households. But the trend has been toward large units: in the year 2000, projects with more than 500 kW_p accounted for just over 10% of new capacity, whereas in 2012 they accounted for nearly half of a much larger annual addition (Fraunhofer ISE 2012). The shift toward larger projects has been the principal reason for rising performance; load factors have also been rising, slowly but steadily. The average irradiance of 1,055 kWh/m^2 of flat surface is boosted to about 1,200 kWh/m^2 by appropriate tilting (30–40 degrees) of the panels, and with a performance ratio of 0.85, the effective annual radiation input is about 1,020 kWh/m^2 (Wirth 2013). That implies an average exploitable flux of about 116 W/m^2.

With an average 11% efficiency, that amounts to an annual electricity generation of about 112 kWh/m^2 and an average power density of more than 12 W_e/m^2 of roof area covered by modules—and all that without making any land claims. Using an average PV generation full load of nearly 970 hours (Fraunhofer ISE 2012) and the exploitable flux of 116 W/m^2 confirms this density ($970 \times 116 = 112.5$ kWh/m^2). In more sunny locations the power densities of rooftop PV installations will be proportionately higher.

McKay's (2013) maxima for US rooftop installations are 20.69 W_e/m^2 for a 390-kW project in Hawaii and 17 W_e/m^2 for an 830-kW installation in California.

The practical maximum capacity of a roof-based PV module is easy to calculate for a particular house or a commercial or industrial enterprise, but nationwide estimates are not that easy to quantify. Many roofs are obviously either poorly suited or entirely unsuitable for such installations because of excessive pitch (greater than 40 degrees; on the other hand, the slope should be at least 15 degrees for self-cleaning), suboptimal orientation, or shading by surrounding buildings or trees, and many roofs are unavailable because of the presence of heating, air conditioning, and ventilation equipment. Denholm and Margolis (2008b) cite Navigant Consulting data that assume 22% availability of roof area for residential buildings in cool climates and 27% in warm and arid climates (the difference owing to reduced tree shading), while for commercial buildings the means are 60% in warm climates and 65% in cooler climates.

A German study (based on sampling of countryside, village, and suburban homes in Bavaria) assumed that 80% of the area of all sloping south-facing roofs of houses and 50% of the area of flat roofs of industrial buildings were available for PV installations (Lödl et al. 2010). This resulted in a PV potential of 8.7 kW_p for suburban houses with an average footprint of 116 m^2 and 12.5 kW_p for village houses with a built area of 167 m^2. Based on that sample, the Bavarian PV rooftop potential was put at 25.3 GW_p, the nationwide total was estimated at 161 GW_p, and the authors cite two other estimates of total German rooftop PV potential at 53–116 GW_p and at 130 GW_p. For comparison, German rooftop capacity was nearly 18 GW_p in 2011, or one-third of the lowest potential estimate.

Although differences in house sizes and in residential and industrial population densities mean that national rates are not readily transferable, it is interesting to note that the German mean amounts to 75 W_p/m^2 of building footprint, and so, at least in countries with similar residential/industrial patterns, that rate might be used to approximate a national rooftop PV potential by using a much more readily available total of built-up area. An even more interesting recent study considered the degree to which the loss of nuclear generation in post-Fukushima Japan could be replaced by rooftop PV-based electricity generation in Tokyo (Stoll, Smith, and Deinert 2013).

The study used data from 34 years of solar irradiance for the Tokyo metropolitan area (averaging 154 W/m^2). A satellite-based analysis of the area available for rooftop greening in the city (flat and free of obstruction) found a total of 50.69 km^2 (assumed to be available for PV panels) and an adjusted total of 10.03 km^2 for sloped house roofs. The suitable rooftop area of Tokyo's 23 wards was put at 64.28 km^2, and the total was 204.05 km^2 for Kanto, the region surrounding the city that is supplied by the Tokyo Electric Power Corporation (TEPCO): that area could support 43.1 GW$_p$ of PV capacity, and if its generation were coupled with the region's existing pumped storage of 7.28 GW, the combined system could provide 4.8 GW 91% of the time. But this was a theoretical exercise that assumed an unrealistically high installation coverage.

A further step toward landless solar electricity generation is the installation of PV walls. Thin-film PV cells made of copper indium gallium selenide can be laminated directly into walls (and, obviously, into roofing materials), and as their efficiency rises to rival that of the silicon-based cells, it will become more appealing to embed them into the south-facing walls of new buildings. Tall buildings offer the greatest opportunities but also experience significant reductions in insolation owing to shading, and wall PV installations would have suboptimal angles of irradiation. These realities are illustrated by a study of a pioneering PV wall installation, an integrated curtain wall facade that spans 12 floors of the lower part of the Solaire Building in New York, built in 2004 (Perez et al. 2012).

The PV array of monocrystalline silicon cells covers 153.5 m^2, has a peak capacity of 11.3 kW, and faces the Hudson River waterfront, and hence its azimuth is 275 degrees. That is hardly an optimal orientation and one whose only advantage is a largely unobstructed exposure except for shading by a few trees. The building's wall receives 766 kWh/m^2. The unshaded rate would be 822 kWh/m^2, a horizontal area in the same location would get 1,430 kWh/m^2, and the tilted south-facing surface would receive 1,615 kWh/m^2, or 2.1 times as much as the actual wall installation, whose annual electricity generation was only 5,560 kWh/year (635 W), resulting in a power density of just 4.1 W$_e$/m^2.

Concentrating Solar Power

Concentrating solar power (CSP) stations use tracking (computer-controlled) parabolic mirrors (heliostats) to reflect and concentrate radiation on a central receiver placed on a high tower. The concentrated radiation is then used to heat a transfer fluid (molten salt, whose temperatures reach up to 650°C), which then heats steam to power a turbogenerator. This technique has three obvious advantages when compared to PV plants: it can achieve higher conversion efficiencies; it can be used in a dual arrangement with fossil fuel or wood to generate steam during the night or during periods of higher demand; and a part of the peak heat flux can be stored in order to generate electricity at night or during periods of low irradiation, with molten salt as the best storage medium (Azcárraga 2013). Despite these advantages, the typical power densities of CPS-based electricity generation are not superior to those of PV-based electricity generation.

Solar One, the pioneering solar tower project designed by the US Department of Energy and located east of Barstow, California, generated electricity between 1982 and 1986 (CSP World 2012). Its field of 1,818 tracking heliostats covered 72,650 m^2. The project was reopened in 1995 as Solar Two, with added heliostats and with the use of molten salt heat storage to smooth the fluctuating irradiation. Solar Two generated 17.5 GWh/year (an average rate of 2 MW) from the total area of 82,750 m^2 of heliostats (USDOE 1998). The average power density of Solar Two was about 24 W_e/m^2, but after only four years the plant was shut down, and in 2009 the tower was demolished and all heliostats were removed.

Europe's first commercial solar tower project, Spain's PS (Planta Solar) 10, completed by Abengoa Solar in Sanlúcar la Mayor in 2007, is rated at 11 MW_p and generates 24.3 GWh/yr, that is, 87.5 TJ/year at a rate of 2.77 MW (Abengoa Solar 2013; fig. 3.3). At 25%, its capacity factor is fairly high. Its heliostats occupy 74,880 m^2 (624 × 120 m^2), and the entire site is about 65 ha. This translates to a power density of about 37 W_e/m^2 (for heliostats), and to a bit more than 4 W_e/m^2 for the plant's total area; the latter rate is very similar to the performance of PV-based plants. PS20 (in operation since 2009) is rated at 20 MW_p. It generates 48.6 GWh/year (175 TJ/year at a mean rate of 5.55 MW) and has a slightly higher capacity factor of nearly 28%. With mirrors covering 150,600 m^2, the project's heliostat power density is 36.85 W_e/m^2, almost identical to that of PS10, but at 6 W_e/m^2 the rate for the entire site (about 90 ha) is nearly 50% higher.

The world's largest CSP project is Ivanpah Solar Electric Generating System (SEGS) in the Mojave Desert in San Bernardino County, California, a site with an exceptionally high annual irradiation of 2,717 kWh/m^2

Figure 3.3
Abengoa central solar power plant. PA Pundits—International.

(310 W/m^2). The project is owned by NRG Energy, Google, and Bright-Source Energy, and the three fields have a total of 173,500 heliostats covering 260 ha (the entire project area is 1,400 ha) serving three 138-m-tall towers. The project's total gross installed capacity is 392 MW$_p$ and the expected annual generation is 1.079 TWh, that is, an average rate of 123.2 MW (BrightSource 2013). These specifications prorate to power densities of 47.4 W$_e$/m^2 for the heliostat area and 8.8 W$_e$/m^2 for the entire project area. The first rate is far higher than the power densities of the best currently operating PV facilities. No stunning improvements are foreseen for CSP efficiencies. This makes it safe to conclude that optimally located solar concentrating plants will generate electricity with power densities of 40–50 W$_e$/m^2 of their large heliostat fields and with rates no higher than 10 W$_e$/m^2 of their entire site area.

There is yet another choice to concentrate solar power, not by focusing sunlight onto a single point but by deploying large numbers of Fresnel lenses to concentrate sunlight (raising its intensity by two to three orders of magnitude) onto individual mulitjunction PV cells. The best efficiencies of these expensive cells are now in excess of 40%. Fthenakis and Kim (2013) prepared a life-cycle assessment of such a system, the Amonix 7700 high-concentration PV array in Phoenix, Arizona. This massive tracking unit has an area of 267 m^2, and its installed peak capacity of 53 kW was expected to rise to 62 kW with improvements in the optical bath and better lens tuning. These specifications translate to a power density of 231.2 W$_e$/m^2, an order of magnitude higher than that for nonconcentrating PV installations.

Potential Gains and Limits

Increased power densities will come with further gradual improvements in PV conversion efficiencies. News of such gains come regularly: in 2014 the record was held by Sharp's concentrator triple-junction compound solar cell, which used Fresnel lenses to concentrate radiation onto a layered cell made from, from the top, InGaP, GaAs, and InGaAs on a silicon substrate, at 44.4% (Sharp 2013). And the record for a commercially available multi-junction cell with modular design (actually four cells stacked one on top of another) designed by Semprius (North Carolina), reached 35.5% (Semprius 2013), and the company believes that the efficiency could eventually rise to 50%. Another innovation that could boost photovoltaic's conversion rates involves cell coatings made from organic and inorganic semiconductors that maximize the harvesting of radiation (Tabachnyk et al. 2014).

But in the near terma advances in large-scale electricity generation will come from cheaper modules, not just silicon-based but also increasingly made from inexpensive thin-film perovskites (calcium titanium oxide, $CaTiO_3$). In light of the history of continuing efficiency gains in PV-based electricity generation it is only a matter of time until the best annual power densities of PV conversions in large stations commonly surpass 10 W$_e$/m^2, and it is not unrealistic to think that in 20–30 years, solar plants in the sunniest locations will routinely approach, and surpass, not just 20 but even 30 W$_e$/m^2.

Significant power density gains would be realized with three-dimensional solar energy generation. Bernardi and co-workers (2012) explored such possibilities by modeling and building experimental 3D PV structures (3DPV) that combine absorbers and reflectors in the absence of sun tracking. Their three choices—an open cube, an open parallelepiped twice as tall as the cube, and a tower using slanted panels—could generate power densities that were 2–20 times higher per base area than those of stationary flat PV panels, that is, maximum rates in excess of 100 W$_e$/m^2. In comparison, the gain for a flat panel with dual-axis tracking would be only 30%–80%. Of course, this increased density required a larger cell area (by a factor of 1.5 to four compared to flat panels), but this drawback is more than compensated for by other advantages of the 3D designs: compared to flat stationary panels they can double the hours of peak power generation, and they can greatly reduce seasonal, latitudinal, and weather variations. When combined with inexpensive thin PV films they could open up new possibilities for large-scale PV-based electricity generation.

Another advantage of PV systems is their relatively high safety. Fthenakis and Kim (2011) studied material and energy flows in four commercial PV designs, those of monocrystalline silicon, multicrystalline silicon, ribbon silicon, and cadmium telluride. They concluded that the PV cycle is much safer than conventional energy sources in terms of both statistically expected and possible maximum consequences. At the same time, a German-like mass installation of both rooftop PV units and large multi-megawatt projects is not applicable to all kinds of environments and to all levels of economic development. Germany is not an obvious choice for the world's largest installed PV capacity (the average irradiance over large parts of southern Spain and Italy is nearly twice as high), but the country's combination of advantages will not be replicated anywhere else anytime soon.

Germany has a modern electrical grid and distribution system, diverse manufacturing capabilities, and technical prowess, all of which made it fairly easy to ramp up the production of modules (which was later undercut to a large extent by cheap, subsidized imports from China) and facilitated the production and installation of the needed infrastructure for distributed PV-generated electricity (panel frames, wiring, inverters, meters, and increasingly also storage batteries). Germany's rainy climate provides natural cleansing of modules, and the traditionally fairly high electricity prices have made even early PV-based technologies relatively more competitive than in countries with inexpensive electricity. Also, a large segment of the German population prefers to adopt, and is willing to pay for, for more expensive renewable energy sources. And, of course, generous feed-in tariffs with prices guaranteed for two decades have offered a sure way to make PV-based electricity generation profitable for anybody who has the initial capital investment.

This combination explains why the pace of rooftop PV advances has been much slower in countries where even a few of Germany's advantages are missing: American sensibilities are not as green, and the Americans have always paid much less for their electricity, but a large part of the United States has irradiance twice as high as Germany's, and its high-voltage grid, electricity distribution, manufacturing potential, and technical skills are not inferior. But thanks to high feed-in tariffs, in 2012 Germany had in per capita terms 16 times as much PV capacity as did the United States (Fraunhofer ISE 2012). And replicating Germany's PV achievements would be outright impossible in countries with dodgy grids and unreliable electricity distribution (including a large share of illegal hookups).

It would also be a challenge in nations with a shortage of the technical skills needed to deploy the PV infrastructure for millions of units; in climates where seasonally heavy deposition coats the modules in dust and where wind-driven sand pits module surfaces; in societies where subsidized energy prices have created unrealistic expectations about the cost of the future energy supply and where the abundance of domestic energy resources does not create such urgency as does Germany's high dependence on fossil fuel imports; and in economies where only very few people could afford the initial investment in rooftop PV units, even it was assumed that house roofs were accessible and free for installation: in many densely cities of the Middle East and Asia they are not because they are covered either by illegal structures or by rubbish.

Further, systems considerations dictate that major shares of PV-based electricity generation (between 5% and 15% of the total supply; in 2013 Germany derived 4.7% of its electricity from PV-based generation) will be possible only with greatly expanded storage, a strategy that is now pursued both by Germany and in California. In Germany the program, started in May 2013, is limited to a small PV system of up to 30 kW, and a subsidy is offered for up to 30% of the price of storage systems tied to new or existing PV units. In California the three major utilities will have to buy 1.325 GW of storage capacity by the year 2020.

Wind and Wind-Generated Electricity

Only a tiny fraction of solar energy reaching the Earth goes into energizing the global atmospheric circulation, and hence the aggregate power of exploitable wind is orders of magnitude smaller than that of the planetary irradiance. Still, some recent estimates have concluded that it is many times larger than the global total primary energy supply (TPES). Archer and Jacobson (2005) assessed the global wind power potential at 72 TW (compared to the 2012 TPES of nearly 17 TW) even if only 13% of the Earth's windiest regions were exploited. Lu, McElroy, and Kiviluoma (2009), assuming a larger area and more powerful turbines, put the total land-based potential nearly 75% higher, at 125 TW. Later, Jacobson and Archer (2010) implied that more than 170 PW of wind power are available for extraction in the atmospheric boundary layer region.

Such high totals are in line with the arguments by Roberts and co-workers (2007), who claimed that the power that could be extracted from the planetary jet streams is one to two orders of magnitude greater than that extractable by equal-sized ground-based wind turbines, and proposed that a tethered rotocraft be used to extract this enormous flux, reaching horizontal power densities up to 20 kW/m^2, compared to less than 500 W/m^2 for a large modern ground-based turbine. Miller, Gans, and Kleidon (2011) exposed the fallacy of these high estimates by pointing out that all of them neglected energy conservation, and proceeded to derive a realistic estimate first by tracing the fundamental top-down process of energy transfer.

About 25% of the incoming solar radiation (45 PW) creates pressure differences due to differential heating (Lorenz 1976). About 2% of that total (900 TW) is the maximum available for wind power extraction; half of that total is dissipated in the atmospheric boundary layer (Peixoto and Oort 1992), and 25% of the remainder (112 TW) is dissipated over land, of which no more than 60% (68 TW) could be extracted over nonglaciated terrain. Miller, Gans, and Kleidon (2011) then proceeded to refine this estimate by adding a simple momentum model with reanalysis of wind data and by using climate model simulations. They concluded that the maximum wind power that could be extracted from the atmospheric boundary layer over all nonglaciated land (limited by the rate of its generation in the climate system) is, depending on the estimation approach, as low as 18 TW, and no higher than 68 TW. The lower estimate would prorate to 0.15 W/m^2 of ice-free land, the higher one to 0.57 W/m^2, both being merely theoretical maxima of extractable kinetic energy.

Similarly, Adams and Keith (2013) used a mesoscale model to demonstrate that wind power production is limited to no more than about 1 W/m^2 for any large-scale wind farm with turbines spaced over an area larger than 100 km^2. Obviously, in practice only a very small share of the

Earth's wind energy that could be captured will eventually be converted to electricity by commercially viable turbines. I will assert, with a high degree of confidence, that tethered rotocrafts will not be delivering electricity on a terawatt-hour scale anytime soon, and that the vast continental regions with very low wind speed and with seasonal doldrums will not see any installations of extensive wind farms. Hoogwijk, de Vries, and Turkenburg (2004) used annual wind speed data from the UK's Climate Research Unit to put the economic potential (cutoff at about $1/kWh) at 96 PWh/year (10.96 TW).

The best assessed new total of global wind capacity comes from a study that combined a reanalysis of wind speed data with assumptions about updated wind turbine performance, land suitability factors, and average costs, including those of requisite long-distance transmission (Zhou et al. 2012). Its central assumptions resulted in a total global economic (at less than 9 cents/kWh) wind generation potential of approximately 119.5 PWh/year, or 13.6 TW. Sensitivity analyses show that the estimates depend particularly on assumed wind speed (ranging from −70% to +450%), land suitability (from −55% to +25%), turbine density (from −60% to +80%), and cost and financing options (from −20% to +200%).

The central estimate—13.6 TW, or just 0.11 W/m^2 of ice-free land—is obviously depressed as a result of large continental areas with low wind speeds, particularly common during extended high-pressure spells in the northern hemisphere. In contrast, the power densities for windy coastal and inland regions will be at least an order of magnitude higher. Wind's considerable vertical power densities (Kapitsa's Umov-Poynting vector, explained in the second chapter of this book) translate into relatively low (horizontal) power densities of electricity generation even at the windiest sites. I will illustrate this progression of values by using actual data for one of the larger machines on the market, the Vestas model V90, which has a rated power of 3 MW (Vestas 2013).

The turbine begins generating once wind speed reaches 3.5 m/s, its cut-out is at 25 m/s, and its rated wind speed (v) is 15 m/s. With the rotor diameter of 90 m, the three blades sweep (with the nominal revolution of 16.1 rpm) an area (A) of 6,362 m^2. These ratings (and air density $\rho = 1.2$ kg/m^3) set the maximum power (P) of this Vestas turbine at 12.88 MW:

The Betz limit (0.59; Betz 1926) takes that down to a maximum achievable rating of 7.6 MW, and the actual rated power of 3 MW means that the

Box 3.3
Maximum power of Vestas wind turbine

$P = \frac{1}{2}\rho A v^3$

$= \frac{1}{2} \times 1.2 \text{ kg/m}^3 \times 6{,}362 \text{ m}^2 \times (15 \text{ m/s})^3$

$= 12{,}883{,}050 \text{ W}$

machine has a fairly high correction factor (0.39, owing to performance losses in mechanical energy conversions) and that the Umov-Poynting vector (vertical energy flux) of its electricity generation is 471.5 W/m^2 of the area swept by its blades.

Power Densities
Compared to this fairly high rate, the power densities of wind-driven electricity generation as energy flux prorated per unit of horizontal surface are two orders of magnitude lower because wind turbines must be set apart to avoid excessive wake interference. Turbines must be placed at least three, and better five, turbine diameters apart in the cross wind direction, and at least six and preferably ten diameters in the downwind direction (Hau 2005). The power density of large wind farms with turbines on a square grid is thus 2–3 W_i/m^2. Returning to the Vestas 90, spacing these machines in a large wind farm on a regular rectangular grid five and ten rotor diameters apart would result in a power density of 2.37 W_e/m^2 for the rated performance of 3 MW_i/turbine. The American Wind Energy Organization lists the power densities of more than 70 large-scale (what they call industrial) facilities on its website, with the rates ranging from just below 1 W_i/m^2 to 11.1 W_i/m^2 at an exceptionally windy site (Cassia County, Idaho) and with most densities, as expected, between 2.5 and 4 W_i/m^2 (AWEO 2013).

In reality, turbine configurations in large wind farms include single strings (sometime with gaps), parallel strings (often close to a perfect grid spacing), multiple strings that are not uniformly oriented, and clusters laid out in irregular fashion, dictated by terrain or microscale differences in wind speed (fig. 3.4). Denholm and co-workers (2009) studied all of these configurations belonging to 172 existing and proposed US large-scale wind projects (with installed capacities greater than 20 MW) in all windy regions

Figure 3.4
Layout of a wind farm: 100 1.5-MW turbines in Shiloh, Solano County, California.
© Proehl Studios/Corbis.

of the nation, and their conclusions offer a representative quantification of both typical and extreme land claims of modern wind farms.

Denholm and co-workers (2009) used the land total listed in project applications or associated documentation and found that the average total land claim was 34 ± 22 ha/MW, for a power density of 3 ± 1.7 W_i/m^2 (with the extreme values more than 15 and less than 0.5 W_i/m^2). Plotting power density as a function of wind farm size does not show any pronounced correlation: most projects with an overall capacity of less than 100 MW have densities of 2–6 W_i/m^2. For larger projects (more than 300 MW) the density declines a bit, but its spread narrows to 2–4 W_i/m^2. But all wind power densities cited so far exaggerate the real performance because all were calculated using rated capacities that were not corrected for prevailing capacity factors calculated by using actual annual electricity-generation totals. These rates vary not only among sites but also show significant interannual fluctuations for the same site.

For many early large-scale wind projects of the late 1980s and 1990s, the capacity factors were well below 20%. A detailed examination of the actual record of wind generation in the EU, which has the world's largest concentration of wind power, shows that during the five years between 2003 and 2007, the average capacity factor was less than 21% (Boccard 2009). In 2010 the nationwide means were 25% for the UK and 24% for Spain and Denmark, but only 20% for France and 15% for Germany. In the United States, better siting and better turbine designs have resulted in noticeable long-term gains in average capacity factors. An analysis of 94% of all US projects built between 1983 and 2010 shows the average load factor rising from 25% in 1999 to 33% in 2008, then dropping to about 30% in 2009 and 2010 before rebounding to 33% in 2011 (Wiser and Bolinger 2012).

A year later Wiser and Bolinger (2013) noted that the rate for 2006–2012 (32.1%) was higher than for 2000–2005 (30.3%), but that the trend had not been either as significant or as consistent as expected, and that the 2012 rate was below the peak achieved in 2008, and, most important, the average capacity factors for projects built after 2005 had been stagnant. The explanation is simple: while better turbine designs boosted capacity factors, the locations of many new projects in less consistently windy areas tended to lower it. In 2012 the average wind resource at the height of 80 m was 15% lower than among the turbines built in 1998–1999.

This seemingly irrational location choice makes sense because lower resource quality occurs in locations closer to major markets and readily connected to existing transmission lines—and an inferior capacity factor is compensated for by the savings on high-voltage transmission lines that would be needed to bring electricity from more windy, but more distant, locations. The same consideration has kept Germany's capacity factors quite low. The country has EU's highest wind capacity (about 30% of the total in 2012), much of it in only moderately windy areas, and during the first six months of 2013 German wind turbines had a capacity factor of only 16%, ranging from 11.8% in May to 21.7% in January (Chabot 2013). This poor European performance has been a major reason why all but one of the wind projects rated by Standard & Poor's have fallen from investment grade to speculative grade over time (Standard & Poor's Rating Services 2012).

Taking the most recent US mean of 32% would thus be a proper correction factor for American capacities and would imply—using the previously established average power density of America's large wind farms of 3 ± 1.7

W_i/m^2—that the power density of the country's wind-driven electricity generation is only 0.96 ± 0.54 W_e/m^2. McDonald and co-workers (2009) used a slightly higher capacity factor of 35% when calculating their range of 1.4–1.7 W_e/m^2 for the least and the most compact US projects. For the EU, the actual power densities should be calculated with an average capacity of factor of only about 20% (25% for the windier UK and Spain), lowering the rate to just 0.6–0.75 W_e/m^2. This is an order of magnitude lower than the power densities of solar electricity generation and (as I show later in this chapter) only two to three times as high as the densities of the most productive liquid biofuel industries.

But the comparison must be qualified because of some fundamental differences in the nature of occupancy of the claimed land and the degree of its permanence. Obviously, crops or tree plantations do not allow any other concurrent land use, nor is the area covered by closely spaced PV cells suitable for any other uses. And while the strips between PV arrays in large solar parks are theoretically available for grazing, the coexistence of PV modules and sheep will not be a common occurrence. In contrast, on a large wind farm the land completely excluded from other uses is limited to that covered by turbine pads, transmission infrastructures (including substations), permanent access roads, and service buildings.

The National Renewable Energy Laboratory's wind farm area calculator assumes that this permanent footprint amounts to 0.25 acres per megawatt, that is, 1,000 m^2/MW or 1,000 W/m^2 (NREL 2013). But according to Denholm and co-workers (2009), direct land claims (permanent and temporary during installations and repairs) average 1 ha/MW, of which 0.7 ha/MW reflects temporary claims during construction and 0.3 ha/MW is permanent land occupation. This last rate translates to a power density of 333 W/m^2, or three times as land-intensive as assumed by the NREL's area calculator, but even so it would be less than 1% (3,000 $m^2/340{,}000$ $m^2 = 0.88\%$) of the total area claimed by an average large American wind farm.

Consequently, 99% of land could be devoted to a variety of agricultural (annual or permanent crops), horticultural (flower beds), or silvicultural (tree and shrub nurseries, Christmas tree plantations) uses, or it could be grazed by domestic animals. Unlike in the case of biofuels or hydro energy, the low power densities of wind-driven electricity generation are of concern not because of relatively large areas of land transformed by energy conversion but because they indicate the spatial limits of wind exploitation: we cannot increase the number of vertical-axis turbines as a way to maximize wind-powered electricity generation within a given area.

This means that in smaller-sized countries with limited wind resources it would not take very long before the continuing exploitation of wind had to move to locations of lower wind quality (lower average speed, lower persistence). But in nations with large territories, many high-quality wind sites that are far away from major urban and industrial areas will remain unexploited until the development of requisite transmission lines connects those relatively high-power-density locations with large electricity markets. In the United States, that would mean building more than a score of long-distance lines on a semicontinental scale, and multiple lines connecting the windy Great Plains with the coasts cannot be completed in a matter of years (Smil 2011).

Noise Effects

Finally, there is one spatial consideration besides the requisite machine spacing that limits the location of large wind turbines: they can overtop crops or grazing animals, but they cannot be sited right next to permanent settlements (as even very large solar parks can) because of the turbine noise. The mechanical component of the noise (from gearboxes) has been reduced by better design, so the concern is about the aerodynamic (whooshing) noise produced by the flow of air around the blades and tall towers. Typical maximum noise levels for human exposure are set at 40 dBA sound intensity (integrated over the 20–20,000 Hz band) with a wind speed of 8 m/s at 10 m height. This usually means buffer zones of about 350 m for large wind farms. At that distance the noise will be 35–45 dB, while a busy office may register at 60 dB and a quiet bedroom at 20 dB.

Many jurisdictions do not specify decibel levels as they have simply set standard buffers. Scottish planning policy calls for 2 km between wind farms and the edge of cities and villages; Wales has a 500-m recommendation (Regen 2012). But in 2013, after Milton Keynes Borough Council wanted to set a 1.2-km buffer, the UK's High Court of Justice ruled that local councils could not impose their own buffer zones for new wind projects (Royal Courts of Justice 2013). The United States has some 25,000 zoning jurisdictions, with most setback rules set at the country level (USDOE 2011b) and with some buffers being just 150 m from a property line containing a dwelling while others require more than 1 km. A 63-MW farm (21 3-MW Vestas turbines with a 90-m rotor diameter) with regular five by ten-diameter spacing would require 486 ha (about 13 W_i/m^2), but adding a 1-km noise buffer zone on all sides would increase the claim to 1,786 ha, 3.7 times the spacing claim, and lower the power density to 3.5 W_i/m^2.

What is tolerable and what is objectionable? Large weather variations and terrain specifics mean that long-term measurements of noise from wind turbines are needed before actual impacts can be assessed. Such measurements may show that the urban background noise level (above all from road traffic) may dominate at distances beyond 300 m from a wind turbine (Björkman 2004). But this is hardly universal. As with other controversial topics involving complex mental and physical human responses to environmental disturbances, the findings of wind turbine noise studies span a wide range of outcomes supporting an equally wide range of views, from those who see a new, full-blown "wind turbine syndrome" of health impacts (Pierpont 2009) to those who dismiss the concern as nothing but "a prime example of a contemporary psychogenic illness" (Chapman 2012, 1).

Such a dismissal is not supported by a great deal of accumulating evidence. There is no need to demonstrate any direct physical effect of noise: noise annoyance can act as a mediator that leads to sleep disturbance and mental distress (Bakker et al. 2012). And there are objective assessments of the effect. Nissenbaum, Aramini, and Hanning (2012) studied the effects of wind turbine noise on sleep and health at two sites in Maine and demonstrated that participants living within 1.4 km of a wind farm had worse sleep, were sleepier during the day, and had worse mental component scores than those living farther away. The overall evidence is complex (Roberts and Roberts 2013), but there is no doubt that living too close to large industrial wind turbines can harm human health and that typical symptoms are stress disorder–type diseases acting by indirect pathways (Jeffery, Krogh, and Horner 2013). Dealing with this reality would further dilute the average power density of wind generation from turbines located in more populous regions, including rural areas in many countries in Asia.

At the same time, there are some intriguing possibilities of boosting the typical power densities of wind-driven electricity generation. One of them might be to move away from using the horizontal-axis turbines that dominate the global market to using vertical-axis machines, whose denser spacing can raise the output per unit area. Experimental field tests with six 10-m-tall and 1.2-m-diameter vertical-axis wind turbines with 4.1-m span airfoil blades (a modified version of a commercial model by Windspire Energy) that were spaced 1.65 diameters apart with a footprint of 48.6 m^2 demonstrated opportunities for raising average power densities by extracting energy from adjacent wakes and from above the wind farm (Dabiri 2011). With three turbines rotating around their central shaft clockwise and the other three rotating counterclockwise, daily mean power densities with winds above 3.8 m/s ranged between 21 and 47 W/m^2 at wind speeds above cut-in speed, and between 6 and 30 W/m^2 overall during the three months of testing, an order of magnitude above the power density of horizontal-axis wind turbines.

Another option—one that uses well-developed horizontal-axis turbine designs and that has already been commercially demonstrated in some countries—is to move away from land and set up large offshore wind farms. Power density limitations cannot be avoided (offshore wind turbines still have to be spaced appropriately), but stronger and more reliable offshore winds result in higher capacity factors and remove the common environmental concerns associated with land-based wind electricity generation. Denmark was the first country to build a substantial offshore capacity, with 871 MW installed by the end of 2012; in that year Danish offshore farms had a capacity factor of 44.9% and a lifetime capacity factor of 39.1% (Energi Styrelsen 2013).

But the costs and technical challenges (building longlasting structures in a corrosive environment, the construction of new long-distance, high-voltage transmission lines) are hardly trivial. In 2014 the United States still did not have a single offshore wind farm (although the first ones had been planned during the 1990s), Germany found that its offshore aspirations had become "dramatically problematic" (Dohmen and Jung 2011), and analyses done in both the UK and Germany showed that just connecting an offshore wind farm to the land grid costs more than building an equivalent capacity in gas turbines that can be located virtually anywhere. Even so, the European Wind Energy Association envisages great advances not just for near-shore installations in shallow water but for deep-water turbines far offshore (EWEA 2013).

Box 3.4
Power of a water turbine

$$P = \eta \rho Q g h$$
$$= 0.87 \times 1{,}000 \times 700 \times 9.81 \times 118$$
$$= 704.97 \text{ MW}$$

Water Power and Hydro Generation

The energy flux of water turning a turbine (P) depends on the rate of water flow (Q, in m³/s) and on the hydraulic head (h, in m), the distance through which the water falls before hitting the blades, and the actually delivered power will be also a function of the turbine efficiency (η); water density and the acceleration of gravity remain identical ($\rho = 1$ g/cm³, that is, 1,000 kg/m³, and $g = 9.81$ m/s²). A large, highly efficient (87%) turbine working under a head of 118 m and receiving a water flow of 700 m³/s will generate about 700 MW of electricity:

These numbers are the actual specifications for each of the 20 turbines installed in Itaipu, still the world's largest hydro station in terms of annual electricity generation, located on the Paraná River between Brazil and Paraguay. China's Sanxia (Three Gorges, completed in 2009) Dam has a 60% higher installed capacity (22.5 GW, compared to Itaipu's 14 GW) but a much lower capacity factor (about 50%, compared to Itaipu's rate of close to 80%), and hence it typically delivers only about 85% of Itaipu's power (Chincold 2013; CTGC 2013). With little difference in the high efficiency of modern turbines and with ρ and g constant, the power densities of modern hydroelectricity generation will range widely along the continuum governed by the extremes of Q and h. At one extreme are the projects that rely on massive water flows with small generating heads created by low dams. The Yacyretá project on the Paraná between Paraguay and Argentina is an excellent example of this category: its hydraulic head is just 22 m, but its low dam creates a reservoir of 1,600 km² to supply enough water to install 3.1 GW and to generate 20 TWh/year (that is, 2.29 GW and a high load factor of nearly 75%).

The other extreme is represented by the tall dams pioneered in the Alps and creating high generating heads. The world's tallest dam is the 300-m

Nurek on the Vakhsh in Tajikistan, built during the Soviet era and operating since 1980; an even taller, 335-m dam is considered for an upstream location on the same river, but so far the Rogun project has no financing. In some cases of high-head stations there may be no reservoir at all; these projects merely divert part of a mountain stream's water flow into a steeply falling conduit (underground tunnel or aboveground pipes) that leads it to turbines located far below. And there is yet another kind of river-run station (streaming systems), exemplified by Manitoba Hydro's Limestone project on the Nelson River, where a low dam (hydraulic head at 27.6 m) creates virtually no reservoir but the flow (regulated by three dams upstream) suffices to support an installed capacity of 1.34 GW (Manitoba Hydro 2013a). But, not surprisingly, for the Three Gorges Dam, the world's largest hydroelectric project, both Q and h are substantial.

Rated heads determine the type of turbine uses. Francis turbines can work with both low and high heads (10–300 m), Pelton and Turgo impulse designs are used for high heads (higher than 100 m), and Kaplan machines, with adjustable guide vanes, are installed on projects with low to medium high heads (2–70 m). Rated speeds can be up to 1,500 rpm for Francis and Pelton turbines, half that rate for Kaplan machines; and maximum unit ratings can be in excess of 500 MW. Plant structures (the dam, including the spillway, the powerhouse, and for some projects also ship locks or fish ladders) and associated infrastructures (switchyard, access roads) make almost always only very small claims compared to the land submerged by a reservoir.

Power Densities of Large Hydro Projects

Dams built in the lower reaches of major rivers create large, even enormous, relatively shallow reservoirs. Lake Volta, impounded by the Akosombo Dam in Ghana, covers 8,502 km², or 3.6% of the territory of Ghana, and is the largest man-made reservoir in the world. The Churchill Falls hydroelectric project, on the Churchill River in Labrador, reaches almost 7,000 km² and created the second largest man-made reservoir in the world. The Kuybyshev Reservoir, on the Volga was created by the Zhiguli Dam and is the third largest reservoir by surface area, having a maximum extent of 6,500 km². Lake Volta, created by the Akosombo Dam, is thus nearly as large as Puerto Rico (about 8,900 km²), and the Kuybyshev Reservoir is larger than Brunei (nearly 5,300 km²). The power densities of these projects (leaving aside the

comparatively minor areas required for dams, associated structures, transformers, high-voltage connectors, and approach roads) are almost invari-
ably less than 1 W_i/m^2: Akosomobo, with an installed capacity of 912 MW
and a head of 68.9 m (Volta River Authority 2013), rates just 0.11 W_i/m^2,
and the Churchill Falls project rates less than 0.8 W_e/m^2 (fig. 3.5).

Hydro projects built on the middle and upper reaches of rivers have
power densities an order of magnitude higher. Itaipu rates 10.4 W_i/m^2;
Sayano-Sushenskaya, Russia's largest hydroelectric station, on the Yenisei
(6.721 GW_i), is slightly higher, at 10.8 W_i/m^2. The Grand Coulee Dam,
America's largest project, on the Columbia River (6.809 GW_i), comes to
about 21 W_i/m^2; and China's Sanxia (with a reservoir area of 1,084 km^2) has
an impressively high rate of 20.8 W_i/m^2. In contrast, Egypt's Aswan Dam
impounds a huge reservoir of 5,250 km^2 but has an installed capacity of just
2.1 GW, resulting in a power density two orders of magnitude lower, at a
mere 0.4 W/m^2. Naturally, the highest rates belong to stations situated in
high mountains, whose tall dams impound small but deep reservoirs. The
Grand Dixence Dam in the canton of Valois, Switzerland (with the world's
tallest gravity dam), has an installed capacity of 2 GW and its lake is just
4.04 km^2 (Grand Dixence 2013), implying an extraordinarily high power
density of 512 W_i/m^2, but the station is used for peaking power and generates annually just 2 TWh, that is, an average rate of 228.3 MW but still a
very high power density of 56.5 W_e/m^2.

The world's record-holder would be the Nepali Arun III Hydroelectric
Project dam. It was originally to be built with World Bank support during
the 1990s, and since that time there have been many failed negotiations
and delays (Siwakoti 1994). Arun III would essentially be a run-of-river project with just a small, 50-ha reservoir and an installed capacity of 402 MW,
resulting in a power density of just over 800 W_i/m^2. Goodland (1995) confirmed an expected rise in power density with a higher installed capacity.
The average power density of hydroelectric projects with an installed capacity between 2 and 99 MW was just 0.4 W_i/m^2; for plants with capacities
between 500 and 999 MW it was 1.35 W_i/m^2, and for the world's largest
dams (in excess of 3 GW_i) it surpassed 3 W_i/m^2.

The power densities of actual generation depend on water supply (precipitation) and on competing water uses, above all on flood prevention
downstream and withdrawals for irrigation. As a result, capacity factors
range from less than 20% during dry years in semiarid or arid regions to

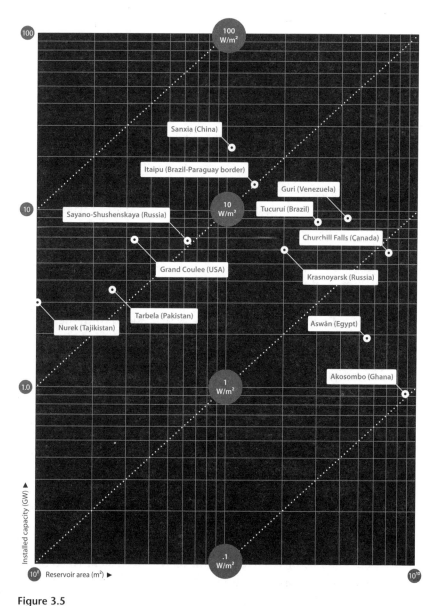

Figure 3.5
Power densities of large hydro stations. Graphed are the installed capacities (GWi) and reservoir areas (in m^2) of the Sanxia, Itaipu, Grand Coulee, Nurek, Tarbela, Sayano-Sushenskaya, Krasnoyarsk, Tucuruí, Gurio, Churchill Falls, Akosombo, and Aswan hydro projects. Carl De Torres Graphic Design.

more than 80% in reliably rainy locations, but commonly are below 50%. For the world's six largest stations these are as follows: as already noted, 50% for Sanxia's and about 80% for Itaipu; Guri (10.2 GW on the Caroní in Venezuela) averages 60%; Tucuruí (8.37 GW on the Tocantins in Brazil) averages 57%; the Grand Coulee Dam just 33%; and Sayano-Sushenskaya about 46%. As a result, Sanxia goes from roughly 20 W_i/m^2 to just over 10 W_e/m^2 and Sayano-Sushenskaya declines to less than 5 W_e/m^3, but the power density of Itaipu remains high, at 8.3 W_e/m^2, while that of the Grand Coulee Dam drops from 21 W_i/m^2 to 7 W_e/m^2. McDonald and co-workers (2009) used an average capacity factor of 44% and put the power densities of the most and the least compact US hydroelectric generating projects at about 7.1 and 1.25 W_e/m^2.

For the stations with the largest reservoirs, the generation densities are less than 0.6 W_e/ m^2 for Churchill Falls and just 0.06 W_e/m^2 as a 10-year mean (2001–2010) for Akosombo (Volta River Authority 2013). Akosombo, affected by large interannual fluctuations in precipitation, demonstrates the need to use longer-term averages. Even for a single decade the annual extremes can depart ±30% from the mean, and hence in 2007 (with only 3.1 GWh generated) Akosombo's power density would have been merely 0.04 W_e/m^2, a rate lower than that even for low-yielding energy crops. Correcting for actual generation is easy, but (as already noted) there is no obvious or generally acceptable way to correct for the multiple uses of many reservoirs.

Pumped Storage

There is yet another challenge in calculating the power densities of water power: how to treat pumped hydroelectric storage (PHES) projects. These stations (the first ones were built in the Swiss and Italian Alps during the 1890s) use cheaper, off-peak (that is, usually nighttime) electricity to pump water from a lower-lying reservoir to a higher reservoir built on adjacent elevated land (fig. 3.6). They can fill it typically in five to six hours, creating hydraulic heads of more than 300 m, with the record difference nearly 700 m (690 m at Chaira in southwestern Bulgaria). During hours of peak demand they serve as a rapidly deployable reserve to provide a virtually instant additional supply for brief periods of time.

PHES projects use reversible pumps or turbines that can switch from pumping to generating mode in just six to ten minutes and that can usually

Figure 3.6
Upper reservoir of Rönkausen pumped storage in Germany. © Hans Blossey/
imagebroker/Corbis.

reach full-capacity generation in a just a few minutes, while turbines synchronized to the grid and spinning on air can go from standby to full load almost instantly. For example, the largest British PHES, Dinorwig, in northern Wales, completed in 1984, with an installed capacity of 1.728 GW, can reach maximum power in just 16 seconds (First Hydro Company 2013). The maximum power of PHES projects is the product of the mass of released water, the generating head, and g (a constant, the acceleration of gravity at 9.81 m/s^2).

For example, Dinorwig can release 7 Mm^3 over a five-hour period to turn its six reversible Francis turbines receiving water from the upper reservoir that is 500 m above the lower storage (First Hydro Company 2013). This translates into a theoretical maximum power of about 1.9 GW, and the plant's actual rated capacity, taking into consideration a roughly 90% conversion efficiency by its large turbines, is about 1.7 GW), and its output is 8.6 GWh.

The same equation can be used to calculate the cost of water pumping. Because it takes 6.5 hours to fill the reservoir, the pumping flow rate is

Box 3.5

Power of Dinorwig pumped storage plant

$P = mgh$

$m = 7 \text{ Mm}^3/(5 \times 3,600) = 388.9 \text{ m}^3/\text{s} \times 1,000 = 388,900 \text{ kg}$

$= 388,900 \text{ kg} \times 9.81 \times 500 = 1.9 \text{ GW}$

$= 1.9 \text{ GW} \times 0.9 = 1.7 \text{ GW} \times 5 = 8.6 \text{ GWh}$

Box 3.6

Energy cost of water pumping

$P = mgh$

$m = 7 \text{ Mm}^3/(6.5 \times 3,600) = 299.1 \text{ m}^3/\text{s} \times 1,000 = 299,100 \text{ kg}$

$= 299,100 \text{ kg} \times 9.81 \times 530 = 1.55 \text{ GW}$

$= 1.55 \text{ GW} \times 1.11 = 1.72 \text{ GW} \times 5 = 8.6 \text{ GWh} \quad 8.6/11.2 = 0.77$

299.1 m^3/s, and because of frictional drag the effective pumping height is 530 m and the efficiency is, once again, about 90%: these specifics result in a pumping cost of 11.2 GWh and an overall efficiency of 77%. PHES operators are willing to lose a substantial share of electricity in order to generate electricity almost instantly at the time of peak demand:

PHES stations are also generally expensive to build, but they remain the only means to store electricity (by converting it to potential power) on a multi-megawatt to gigawatt scale. In 2010 the worldwide capacity of PHES reached 120.7 GW, with Europe having slightly more than a third of the total, Japan a fifth, and the United States nearly a fifth (USEIA 2013b). This means that the worldwide pumped storage capacity was equal to about 13% of all installed hydropower, but the comparison is misleading, for two major reasons. First, hydroelectric power-generating stations and PHES represent two very different modes of generation: hydropower stations generally produce as much electricity as precipitation and water storage will allow, whereas pumped storages generate according to peak demand.

Second, PHES is always a net energy loser, with the most efficient projects consuming about 20% more electricity than they generate. In 2010 the global electricity cost of running PHES facilities amounted to about 22 TWh, and in the United States pumped storages consumed about 5.5 TWh more electricity than they produced (USEIA 2013b). This reality makes it difficult to compare the PHES's performance with other modes of electricity generation. In any case, a project with an installed capacity of 1 GW_e, a reservoir area of about 800 ha, and a load factor of 13% will have an annually averaged power density of just 16 W_e/m^2. In contrast, here are the specifics for a few top projects that show peak power densities up to about 50 times higher.

The Bath County pumped storage station in Virginia, completed in 1985, remains the world's largest pumped storage facility, with 3 GW of installed capacity (Dominion 2013), followed by two 2.4-GW stations in China. The Bath County pumped storage project has an elevation difference of 385 m, the upper reservoir covers 1.07 km^2, and the lower reservoir covers 2.25 km^2 (Dominion 2013). This gives the project a peak power density of just over 900 W_e/m^2, an impressively high rate that could be sustained for only a few hours a day, and hence it is not comparable with the power densities of conventional hydroelectricity stations, which have much higher capacity factors. China's largest pumped storage plant in Guangdong has a head of 535 m, an upper reservoir of 1.2 km^2 , and a lower reservoir of 1.6 km^2 (Chincold 2013). The station uses nighttime electricity from the Dayawan nuclear station to provide peak power to Hong Kong and Guangzhou with a maximum power density of nearly 860 W_e/m^2.

China's second 2.4-GW station, Huizhou, is also in Guangdong province. In 2012 the country had 24 PHES stations with a total capacity of 16.95 GW. Japan's Okutataragi station (1.93 GW) in southern Honshu prefecture and the Ludington pumped storage facility in Michigan (1.872 GW delivered within 30 minutes after startup) complete the world's top five pumped storage facilities. Germany, which would benefit from a higher PHES capacity to accommodate its rising solar and wind electricity generation, had 6.7 GW installed in 2012 (about 5% of total capacity), and despite a common belief that the country's PHES potential is largely exhausted, Steffen (2012) shows that another 4.7 GW could be realized in the coming years. Similarly, more projects are planned for several other EU countries, the United States, and Japan (Deane, Gallachóir, and McKeogh 2010).

Phytomass for Traditional and Modern Uses

In most of Europe, in North America, and in Japan, traditional phytomass fuels—dominated by wood and charcoal, but also including large amounts of crop residues and in some societies also dried dung—were the dominant source of thermal energy until the late nineteenth century. In the rest of the world (including populous China, India, Indonesia, and Brazil), they retained that place for most of the twentieth century. There were only two great exceptions to the dependence on wood in the early modern (1500–1800) world, England-Wales and the Netherlands. England was the first country that accomplished its transition to coal, while the Dutch relied heavily on peat. Warde (2007) suggests that the most likely time when energy from coal combustion surpassed that from the burning of wood was around 1620, and that by 1650 coal provided two-thirds of all primary energy. The Dutch golden age of the seventeenth century was fueled by peat: its per capita consumption was higher than India's average energy supply in the year 2000 (de Zeeuw 1978; Smil 2010b).

In the United States, coal (and the still relatively small flow of crude oil) began to supply more than half of all primary energy by the mid-1880s. In Japan the tipping point came a generation later, while in the USSR it was delayed until the early 1930s and in China until the mid-1960s (Smil 2010b). At the beginning of the twentieth century traditional phytomass fuels provided half of the total primary energy supply; by its end their global harvest had actually doubled as hundreds of millions of poor villagers and many people in smaller towns throughout Asia, Africa, and Latin America continued to rely on phytomass for cooking and heating, but in relative terms those fuels became a marginal part of the total supply, providing no more than 12% by 2010 (Smil 2010b).

In the poorest parts of the world household fuel is still gathered mostly by families, much of it in a way unchanged for millennia, not by cutting trees but by collecting fallen and dry branches of small trees and shrubs. This often requires lengthy walks to the nearest sources of woody phytomass, making it a time-consuming chore that is usually done by women and children. At the same time, in many tropical countries a surprisingly large share of woody phytomass does not come from forests but from roadside and backyard trees and from small groves. And while the increasing availability of kerosene, LPG, and electricity reduced the dependence on crop residues, the burning of cereal straws is still a common practice in many rural areas, particularly in Asia.

In modern countries virtually all phytomass for energy conversions comes from three major sources. The first one is woody phytomass explicitly destined for energy conversions (combustion, gasification), with an increasing share coming from tree plantations of fast-growing species. The second one is a diverse group of residual phytomass that includes those tree parts that do not become merchantable timber or pulp for making paper (bark, chips, wood shavings, sawdust) and crop residues (a category dominated by cereal straws and sugar cane bagasse). The third one, the latest addition to the two well-established sources, is the cultivation of annual or perennial field crops that are converted to liquid biofuels (mostly ethanol and biodiesel) or are gasified.

The power density of phytomass energy use is always very low, an inevitable consequence of the inherently poor efficiency of the conversion chain that starts with solar radiation and ends with actually harvested phytomass. The photosynthetic conversion of select wavelengths of solar radiation to the chemical energy of new phytomass is a remarkable transformation, the foundation of all heterotrophic life and hence also of all human societies and civilizations (Smil 2013a), but one with inherently low efficiency because only a small part of solar energy that is initially converted to new chemical bonds in those plant tissues ends up as harvestable phytomass.

From Solar Radiation to Phytomass

Cannell (1989) offered a representative sequence of efficiencies and conversion losses leading from the insolation to actual wood yield, and I have updated some of his multipliers and recalculated all the steps in terms of power densities for a location with 115 W/m^2 of annual irradiance (typical for the forests in southern Sweden) and with all progressive rates expressed in W/m^2 (rounded to the nearest 0.1).

Conversion of the power density of 0.6 W/m^2 back to annual phytomass yields (assuming an energy density of 19 GJ/t and a specific density of 500 kg/m^3) roughly 10 t/ ha in terms of dry matter. This would be a fairly representative mean for a growing (nonclimax) forest on fairly good soils and receiving adequate precipitation. Indeed, Luyssaert and co-workers (2009) found that the best available model of the net primary productivity (NPP)

Box 3.7
Power densities of a photosynthetic progression

Insolation			115 W/m^2
Photosynthetically active radiation (blue and red light)	× 0.43	=	49.5 W/m^2
Leaf reflectance and transmittance	× 0.85	=	42.0 W/m^2
Quantum efficiency of photosynthesis (theoretical carbohydrate output)	× 0.25	=	10.5 W/m^2
Limited diffusion of CO_2 to chloroplasts	× 0.4	=	4.2 W/m^2
Limits owing to light interception at favorable temperature	× 0.3	=	1.2 W/m^2
Photosynthate reduced by respiration	× 0.5	=	0.6 W/m^2

of European forests (for EU-25) produces an annual mean of 520 ± 75 g C/m^2, that is (assuming that wood is 50% C), 10.4 ± 0.75 t/ha. Another set of estimates puts the mean annual NPP of EU-25 cropland at 646–846 g C/m^2, or roughly 13–17 t/ha (Ciais et al. 2010). Most of the world's terrestrial plants have yields, and hence production power densities, of the same order of magnitude. The NPP (with all values in dry matter) of tropical ecosystems is on the order of 20 t/ha, the means for temperate and boreal forests and woodlands are around 10 t/ha, and the world's cultivated land has a very similar mean.

The global terrestrial NPP is between 110 and 115 Gt/year (Ito 2011; Zhao et al. 2005). If we assume (conservatively) 18 GJ/t, this is at least 2 ZJ, or nearly 63 TW. Prorated over the Earth's ice-free surfaces the average photosynthetic power density would be just 0.4 W/m^2, but a more accurate rate results from also leaving out both hot and cold deserts, which raises the mean power density to about 0.6 W/m^2 (11 t/ha). During periods of most rapid growth, NPP additions could amount to 200 kg/ha a day for the inherently more efficient C_4 plants and up to 150 kg/day for C_3 species. But NPP does not measure actual phytomass harvest; it is a construct of the gross productivity adjusted for respiration. NPP excludes heterotrophic consumption (all preharvest losses to bacterial and fungal infestations, to insects, birds, rodents, and other mammals), as well as losses to inclement weather (wind damage, flooding, drought).

The rate that adjusts for all losses is net ecosystemic production (NEP), and if there are no weather-induced preharvest losses and if the entire aboveground growth is harvested (such as with whole-tree utilization for wood chips or with a forage or silage crop), then the NEP should be identical to the actual harvest. But more often our harvests do not include all the aboveground phytomass. When trees are grown for stemwood (for sawnwood or pulp), the actually harvested net stem growth is only about 20% of NPP and 40% of NEP (Pretzsch 2009). And for cereal crops, which are by far the largest category of agricultural products, the harvested grain is typically no more than 40%–45% of the aboveground phytomass (the rest being cut straw and the remaining stubble). The power density of actual phytomass harvests has changed with time as a result of changes in harvesting methods as well as in the degree of phytomass utilization.

Harvests, Yields, and Power Densities
Studies in tropical and subtropical forests show annual production rates as low as 5–6 t/ha and as much as 12–15 t/ha, whereas in temperate forests the rates are mostly less than 5 t/ha (Bala et al. 2010; Clark et al. 2002; Liu et al. 2003; Odiwe and Muoghalu 2003). Converted by using an average of 18 MJ/kg, these rates would translate to 90–270 GJ/ha in the tropics and subtropics and mostly to less than 80 GJ/ha elsewhere, or between roughly 0.25 and 0.85 W/m². Many litterfall studies also show the composition of the fallen phytomass: leaves (or needles) typically account for 65%–75% of the total mass, reproductive parts (flowers, seeds, nuts) for 5%–10%, and coarse woody debris for only 15%–25% of the total, or as little as 750 kg and as much as 3.75 t/ha a year. Only the last part of litterfall, coarse woody debris, is usually collected, and the annual power densities of collectable woody phytomass would be no higher than 0.04–0.20 W/m².

But in some extreme circumstances virtually the entire litterfall has been harvested in regions with severe fuel shortages, particularly in long-ago deforested areas of North China, where peasants used to sweep up every piece of plant litter, including leaves, needles, and small twigs, shed by remaining wood groves and carry them in baskets on their backs to their house to stir-fry their meager meals; this practice could be seen in North China as recently as during the 1980s. But because the litter yield in those low-density, low-productivity groves was no more than a 3–4 t/ha, even its complete harvest would prorate to less than 0.2 W/m². The lowest woodfuel collection rates have been recorded in the poorest arid parts of Africa; for example, in Eritrea's scrublands and wooded grasslands the annual yield is as low as 30–50 kg/ha of air-dried (15 GJ/t) wood (Arayal 1999). Even the latter rate is just 5 g of woody phytomass per square meter, just a tiny twig and a minuscule power density of 0.0024 W/m^2.

Cutting down mature virgin forests or older succession growth to supply towns and growing cities tapped much richer stores of phytomass but did not proceed with much higher power densities. Clearing a rich temperate primary forest could yield 400 t/ha, but as such a harvest could not be repeated (even after 80 years a secondary growth would yield at least 20% less), a properly prorated annual power density mean was no more than about 0.25 W/m^2. Large cities located in temperate regions—such as China's northern capitals Xian and Beijing, London, and Paris—had considerable demand for wood, a function not only of their climate and size but also of the wasteful combustion in open fireplaces (more efficient enclosed stoves came relatively late even in Europe) and the similarly wasteful conversion of wood into charcoal, the preferred (smokeless) fuel for heating.

All preindustrial averages of per capita fuelwood consumption are just approximations, but they suffice to illustrate annual urban claims. My estimate for the early Roman Empire is no more than 600 kg/capita (Smil 2010c). In London of the early fourteenth century it was 1.5 t/capita (Galloway, Keene, and Murphy 1996). During the eighteenth century a typical German mean was close to 3.5 t/capita (Sieferle 2001). In 1830 it was about 4.5 t/capita in Austria (Krausmann and Haberl 2002). In the forest-rich United States the mean was nearly 6 t/capita (roughly split between household and industrial uses) during the 1850s (Schurr and Netschert 1960), but in Paris it declined from nearly 1 t/capita in 1815 to less than 300 kg/capita by 1850 (Clout 1983). Even if the average consumption were only 1 t/capita, a preindustrial city of million people would have needed every year wood harvested from about 250,000 ha, the equivalent of a square with sides of 50 km.

The prevailing power densities were further reduced by the conversion of wood to cleaner-burning but inefficiently produced charcoal. As already explained, English charcoal-making during the eighteenth century—when the charcoal to wood ratio was 1:5 in mass and (assuming 29 GJ/t of charcoal and 19 GJ/t of wood) about 1:3.3 in energy terms—operated with a power density of less than 0.1 (just 0.07) W/m^2. Iron-makers located in forested regions (resulting, inevitably, in extensive local deforestation), but cities had to import charcoal from increasingly distant sources, even though the easily crushable fuel is not suited for long-distance transport.

An eighteenth-century city consuming roughly half of its fuel demand as fuelwood and half as charcoal would draw on an energy supply produced with power densities of less than 0.15 W/m^2, and for every 100,000 of its inhabitants its annual fuel supply would have required wood harvests from about 40,000 ha of forested land. The combination of expanding areas of deforestation in the vicinity of large cities and the high cost of land transport was an important factor limiting the size of preindustrial cities that had to be supplied by wood delivered by heavy carts (pulled by oxen or horses) or on the backs of donkeys or camels (a common practice in pre-1900 Beijing) from increasing distances.

Harvests of crop residues had much lower power densities than those of woody phytomass, in temperate regions almost always less than 0.05 W/m^2 (less than 1 t/ha). But the two kinds of biofuels are not in the same category, and hence the comparison is flawed: wood is the targeted harvest, while crop residues are just by-products of harvesting cereal, leguminous, oil, sugar, or fiber crops. Moreover, straw and stalks and leaves have always had many more valuable uses than to be burned inefficiently in small household stoves. Such uses range from animal feed to substrates for mushroom cultivation, and the best choice might be to incorporate most, even all, residues into soils to replenish organic matter, retain moisture, and help prevent erosion (Smil 2013a).

Some modern harvests of wood for energy still come from natural forests (mostly from secondary and tertiary growth), but the trend has been to cultivate the trees. The planting of fast-growing tree species grown in short rotations began during the 1960s, mostly with poplar clones (Dickmann 2006). These plantings are dense (some with more than 50,000 stems/ha, to be thinned later) and are harvested after four to six years (rarely, more than 10 years), and their harvest is followed either by coppiced growth or by replanting (West 2013). The recent interest in biomass energy has led to a wave of experiments with fast-growing tree species and to reports of some extraordinarily high yields, even in excess of 50 t/ha. Such claims are often based on a simplistic extrapolation of optimally tended small experimental plots. Care must also be taken to properly convert volumetric yield reports (a common practice in forestry) to mass and energy equivalents.

The specific density of commonly exploited species ranges from less than 350 kg/m^3 for some firs to 500–550 kg/m^3 for maples, ashes, elms, and oaks. Densities change with age, and vary even among closely related species: for example, young fast-growing plantation poplars in Washington state had a density of 370 kg/m^3 during the first three years of growth but averaged 450 kg/m^3 six years later (DeBell et al. 2002), while diverse families of loblolly pine have densities ranging between 440 and nearly 510 kg/m^3 (Belonger, McKeand, and Jett 1997). The UN Food and Agriculture Organization uses an average density of 500 kg/m^3 for conifers and 600 kg/m^3 for leafy trees. The energy density (absolutely dry matter) of commonly harvested trees is mostly between 16 and 19 MJ/kg.

Leaving some dubious extreme claims aside, properly conducted recent experiments—including those by Rao, Joseph, and Sreemannarayana (2000), Hytönen (2008), and Sarlls and Oladosu (2010)—and many recent studies of hybrid poplars (Di Matteo et al. 2012; Klasnja et al. 2003; Paris et al. 2011; Truax et al. 2012) confirm the long-standing knowledge of typical yields in tree plantations (Cannell 1989; Dickmann 2006; Mead 2005). In most temperate settings with natural growth or with moderate fertilization (and perhaps some supplemental irrigation), fast-growing large-scale plantings of pines, acacias, poplars, or willows will yield (depending on climate, soils, cultivars, and inputs) between 5 and 15 t/ha (fig. 3.7). In the subtropics and tropics, commonly cultivated trees (different species of *Eucalyptus* and *Acacia, Leucaena, Pinus,* and *Dalbergia*) will yield 20–25 t/ha (ITTO 2009). If a harvested dry mass of 19 GJ/t is assumed, the temperate rates imply power densities of 0.3–0.9 W/m^2, the tropical ones range between 1.2 and 1.5 W/m^2.

Mechanical harvesters (wheeled or tracked, first developed in Scandinavia during the 1970s and 1980s) perform the entire felling and sawing sequence, from cutting down a tree near the ground to removing its branches and bucking it. Stumps, branches, and treetops remain on-site to recycle nutrients. There is also an option to harvest entire trees (leaving only stumps) by using a sequence of a feller-buncher, chipper, and truck loader, which obviously leads to the recycling only of the nutrients in the stumps. Wood chips are burned (sometimes after preliminary drying) to

Figure 3.7
Pine (*Pinus taeda*) plantation in the United States. USDA.

produce steam for electricity generation, or for both heat and electricity in cogeneration plants.

The combustion of woody phytomass is done most efficiently (close to 90%) in circulating or bubbling fluidized bed boilers (Khan et al. 2009). Gasification (in low- or high-pressure gasifiers) converts as much as 80%–85% of energy feed into a gas whose composition is dominated by CO and H_2 and whose energy density is, depending on the procedure used, as low as 5.4 MJ/m^3 or as high as 17 MJ/m^3 (Worley and Yale 2012). When phytomass-derived gas is used in engines or turbines to generate electricity, the overall efficiency of the sequence could be as high as 35%–40%. Converting wood chips to methanol (CH_3OH) can be done with efficiencies of up to 70% (Methanol Institute 2013).

Typical yields of Brazilian eucalyptus plantations rose from about 12 t/ha in 1980 to 21 t/ha in 2011 (CNI 2012), but sustainable yields are much lower in other tropical regions, usually less than 10 t/ha in Africa, India, and Southeast Asia (ITTO 2009). Even when wood from a highly productive tropical plantation (20 t/ha, 1.2 W/m^2) is burned, maximum power densities would be close to 1.1 W/m^2 for heat generation, 1 W/m^2 for

Box 3.8

Power density of wood burning

10 t/ha × 19 GJ = 190 GJ/ha

190 GJ/31.5 Ms = 6,032 W

6,032 W/10,000 m^2 = 0.6 W/m^2

Box 3.9

Land claimed by a tree plantation

1 GW × 0.7 = 700 MW

700 MW/0.35 = 2 GW

2 GW/0.6 W/m^2 = 3.33 Gm^2 (333,000 ha)

$\sqrt{3.33}$ Gm^2 = 57,735 m

gasification, around 0.8 W/m^2 for methanol production, and less than 0.5 W/m^2 for gas used for electricity generation. The power densities of methanol production and electricity generation based on woody phytomass harvested in tree plantations in temperate climates would be, in most cases, no more than half of the above values. A specific example shows what a temperate-climate harvest of 10 t/ha would imply for generating electricity by burning wood chips in a large power plant. Even with an average energy density of 19 GJ/t the plantation would yield no more than 190 GJ/ha, resulting in a harvest power density of 0.6 W/m^2.

To supply a 1-GW$_e$ wood-fired power plant operating with a capacity factor of 70% and a conversion efficiency of 35% would require an annual harvest of about 330,000 ha of fast-growing tree plantation, the equivalent of a square with sides of nearly 58 km.

The total area needed by a wood-fired electricity-generating plant is quite negligible when compared to a large area of land claimed by phytomass production: even if the generating station and its associated structures (fuel storage, switchyard, office and maintenance buildings, access roads) were to occupy 10 ha (a square with sides of 316 m) it would still be only 0.003% of the land required to grow trees. Obviously, the very low power densities of wood-based electricity generation prevent it from becoming anything but a very minor contributor to the overall supply, with plants usually burning wood waste generated by wood-processing industries.

If only 10% of the US electricity generated in 2012 (that is, 405 TWh, or 1.46 EJ) had to be produced by burning wood, then (with an average 35% conversion efficiency) the country would require about 4.17 EJ (about 132 GW) of wood chips. With an average power density of 0.6 W/m^2, this would claim about 220,000 km^2 of wood plantations, or nearly as much land as all of Idaho or Utah. This calculation also shows why even major yield increases (thanks to better hybrids or to entirely new transgenic trees) would not make a fundamental difference as far as the overall land claims are concerned. Even with a doubling of the assumed 10 t/ha mean (not likely on a large scale in temperate regions for decades to come), a wood-fueled plant with a 1-GW$_e$ capacity would still need annual harvests of all aboveground phytomass grown on a plantation of nearly 170,000 ha, a square with sides of 40 km.

Liquid Biofuels

The power densities of producing liquid fuels from agricultural crops are even lower, although some species saw impressive gains in yields during the second half of the twentieth century. Before the introduction of hybrid varieties the US nationwide corn yield averaged just 1.5 t/ha during the mid-1930s. By 1975 it was 5.4 t/ha, in 2000 it reached 8.6 t/ha, and it peaked at 10.3 t/ha in 2009 (FAO 2014). Similarly, better cultivars and better agronomic management of Brazilian sugar cane helped lift the crop's average yields from 43 t/ha in 1960 to 71 t/ha in 2012 (FAO 2014). By far the most important process used to convert these crops to fuel has been the production of ethanol, and it too has improved over time: for example, Brazil's ethanol yield increased from 59.2 L/t of sugar cane in 1975 to 80.4 L/t in 2008 (Cortez 2011).

Experiments with automotive ethanol predate World War II, but modern large-scale ethanol production began in Brazil with the country's sugar cane–based ProÁlcool program in 1975, and in the United States with corn-based efforts in 1980 (Basso, Basso, and Rocha 2011; Solomon, Barnes, and Halvorsen 2007). By 2010 the two programs had expanded to, respectively, about 50 and 20 GL a year. Brazilian production has been stagnating since 2008 (Angelo 2012), and the supply of the US corn-based ethanol is unlikely

Box 3.10
Power density of US ethanol production

Grain corn yield (15% moisture)	10 t/ha
Moisture adjustment	$10 \times 0.85 = 8.5$ t/ha of dry grain
Ethanol yield	$8,500 \times 0.36 = 3,060$ kg/ha
Energy yield	$3,060 \times 26.7$ MJ/kg = 81.7 GJ/ha
Power density	81.7 GJ/ha = 2,590 W/10,000 = 0.26 W/m2

to grow beyond the 9%–10% share of the automotive market it has reached, thanks to production mandates imposed by the US Congress.

America's ethanol producers have been averaging 2.8 gallons per bushel of wet corn (ACM 2013); in metric units this would be 0.38 kg/kg of dry grain, but Patzek (2006) showed that the theoretical yield of ethanol from corn starch (making up 66.2% of the grain) is 0.364 kg/kg of dry corn. This discrepancy is explained by the industry's practice of counting gasoline denaturant (5% by volume, 8% by energy content) as ethanol. Using the yield of 0.36 kg of ethanol (energy density of 26.7 MJ/kg) per kilogram of dry grain and assuming an average US harvest of 10 t/ha results in a power density of 0.26 W/m^2 for US corn-based ethanol.

Sugar cane is a perennial grass, but because its yields decline with successive cuttings the standard Brazilian practice is to have five harvests, followed by replanting (SugarCane.org 2013). According to Crago and co-workers (2010), for the first harvest the ethanol yield is 10,235 L/ha, diminishing to 5,636 L/ha with the fifth harvest: the average is 6,134.4 L/ha, or 0.41 W/m^2, nearly twice as high as for US corn-based ethanol. Sugar cane ethanol has other advantages: the land used to grow the grass is not usually in competition with land used for food crops; the plant's endophytic nitrogen-fixing bacteria eliminate the need for nitrogen fertilizers; and in Brazil's climate, there is no need for irrigation. The production costs are thus cheaper than for the US corn-based ethanol, but the relative competitiveness is changed by including the costs of transporting cane ethanol and the by-product (distiller's grain, corn oil) credits of corn ethanol (Crago et al. 2010).

The conversion of oil extracted from the seeds of oil plants to biodiesel is done by transesterification, that is, by reacting triglycerides in plant oils with alcohol (ethanol or methanol) in the presence of a base catalyst (Gerpen 2005). The transesterification process converts up to 97% of oil into biodiesel, and rapeseeds contain about 40% oil, which means that nearly 39% of the crop yield can end up as fuel. The crop yields mostly between 2 and 3.5 t/ha (the EU-27 mean is about 2.5 t/ha), but the average is 4 t/ha in the Netherlands. The Dutch biodiesel yield thus averages close to 1.5 t/ha, and its energy equivalent (with 37.8 GJ/t) of 56.7 GJ/ha translates to 0.18 W/m^2. By contrast, the best rapeseed-based performance by a Spanish company was 0.22 W/m^2 (González-García, García-Rey, and Hospido 2013).

The average EU yield translates to only 0.12 W/m^2, and, as in the case of the US corn-based ethanol, it is obvious that this low power density alone precludes this fuel ever supplying a significant share of the EU's large diesel demand. The rapeseed required to cover the EU-27 diesel demand of roughly 260 GW would have to be planted on nearly 220 Mha, while the EU-27 arable land adds up to only about 103 Mha (Eurostat 2012). In any case, owing to the high energy cost of farming and processing inputs, about a quarter of the EU's area could produce rapeseed biodiesel only with a net energy loss (Firrisa 2011). The biofuel with the greatest promise—but with a constantly deferred large-scale production—is cellulosic ethanol made by enzymatic hydrolysis of any phytomass high in cellulose and hemicellulose (any woody material, cereal straws, and intensively cultivated high-yielding grasses).

The latter group includes switchgrass (*Panicum virginatum*), a North American native tolerant of summer heat; reed canary grass (*Phalaris arundinacea*), an up to 2-m-high creeping C_3 species from temperate Eurasia and North America; the giant reed (*Arundo donax*), a common Eurasian rhizomatous plant that can grow up to 8–9 m; and miscanthus (*Miscanthus giganteus*), originally an Asian C_4 species whose rapid growth can reach 3–4 m (4F CROPS 2011; Singh 2013). In experimental plots and in small field settings these grasses yield mostly between 10 and 20 t/ha, and dry matter yields of up to 50 t/ha have been reported for reeds, but high yields could not be maintained in large-scale plantings without adequate fertilization and necessary irrigation.

The potential of cellulosic ethanol has been chronically overestimated: completion dates of large commercial plants have been slipping for years, production costs remain uncompetitive, and the conversion of hemicellulose, which makes up 25%–36% of grass tissues (Lee et al. 2007), remains a challenge. Not surprisingly, even as the first large-scale US cellulosic ethanol plant began operating in 2014, the overall appraisal of the fuel's prospects was titled "Cellulosic ethanol fights for life" (Peplow 2014). And no future technical improvements can change the fundamentals. Even when very high yields (15 t/ha of dry matter) and an average conversion of 330 L of ethanol/t of grass are assumed (Schmer et al. 2008), the power density of cellulosic ethanol would be about 0.4 W/m^2, no higher than that of Brazilian ethanol made from sugar cane, while the NREL's design of a thermochemical pathway by indirect gasification and mixed alcohol synthesis assumes a yield of 8.2 GJ/t of dry feedstock (Dutta et al. 2011), and that would translate to just 0.26 W/m^2 for production based on harvesting 10 t of grain corn stover.

Finally, the power density of biogas generation, an energy conversion technique that has been pioneered in rural areas of China and India to turn organic waste into a low-energy-density gas for household use. The conversion is now used on a large commercial scale to produce gas for subsequent electricity generation. Among the affluent countries, Germany has by far the most extensive national biogas program: in 2011 it had 7,000 biogas plants with an installed capacity of about 2.7 GW_e, and their feedstock was about equally divided between energy crops and livestock excrements (FNR 2012). Because of its very high water content, cattle and pig slurry generates only a small share of biogas, with most of it coming from the fermentation of phytomass. German data show an equivalent of about 100 m^3 CH_4/t of fresh corn silage (FNR 2012), and with 50 t/ha this would produce 5,000 m^3 of methane, or 0.6 W/m^2, and after conversion to electricity (producing about 18.5 MWh) the final power density would be slightly above 0.2 W/m^2.

Geothermal Energy

The sources of the Earth's huge heat flux are yet to be accurately apportioned: they include basal cooling of the Earth's primordially hot core and, above all, heat-producing isotopes of ^{235}U, ^{238}U, ^{232}Th, and ^{40}K in the crust (Murthy, van Westrenen, and Fei 2003). At the ocean bottom the heat flux is as low as 38 mW/m^2 through very old (at least 200 million years) floors to more than 250 mW/m^2 for ocean floors younger than 4 million years.

On land, the youngest crust has an average heat flow of 77 mW/m², the oldest continental shields average less than 45 mW/m², and the weighted global mean is 87 mW/m² (Pollack, Hurter, and Johnston 1993).

Typical terrestrial heat flows are an order of magnitude lower than the flows commonly produced by the photosynthesis of natural ecosystems, too low to be converted to useful energy at the Earth's surface. Drilling gets to progressively higher temperatures: the geothermal gradient (the rate of temperature increase with depth) is usually 25–30°C/km, but the extremes range from less than 15°C/km to more than 50°C/km in volcanic and tectonically active regions, particularly along the plate conversion and subduction zones surrounding the Pacific Ocean (Smil 2008). In selected locations in these regions, hot (above 100°C) pressurized water can be tapped by drilling often no deeper than 2 km and used (directly as steam or indirectly in binary systems where another liquid is heated by hot water in an exchanger) to produce electricity in turbogenerators (Dickson and Fanelli 2004).

In vast continental regions, geothermal surface and near-surface fluxes are too small and their temperature is too low for rewarding conversion to electricity, but in many locations these flows can supply significant volumes of hot water for household, commercial, and industrial uses, and shallow wells can be used for heat pumps. Of course, high-temperature water can be recovered anywhere after drilling sufficiently deep into the Earth's crust, but in regions with a normal geothermal gradient it would require wells at least 7 km deep, where injected water would contact dry hot rocks. Such enhanced geothermal systems (EGS) would be the most intensive way of geothermal capture as a set of deep wells reaching hot rocks would be connected by circulating injected water that would be heated to more than 200°C and withdrawn for electricity generation (Tester et al. 2006).

The commercial exploitation of this option is yet to come, and all existing geothermal plants use hot water withdrawn from wells that are relatively shallow (less than 1 km at Italy's Larderello and New Zealand's Wairakei) or of a medium depth (2–2.5 km in Iceland and California); these wells are sometimes recharged not only with fresh surface water but also with treated wastewater. Geothermal electricity generation began at Italy's Larderello field in 1902; New Zealand's Wairakei came only in 1958, followed by California's Geysers in 1960 and Mexico's Cerro Prieto in 1970. Most of these pioneering installations were gradually enlarged, and other countries began to develop their considerable geothermal potential. By the year 2010 the worldwide geothermal electricity-generating capacity had reached 10.7 GW, with nearly 3.1 GW installed in the United States, 1.9 GW in the Philippines, 1.2 GW in Indonesia, 958 MW in Mexico, 843 MW in Italy, 638 MW in New Zealand, and 575 MW in Iceland. The total annual generation was about 67 TWh, resulting in a capacity factor of just over 70% (Bertani 2010; IGA 2013b).

Wells in geothermal fields could be spread over areas of 5–10 km^2, but (much as in the case of hydrocarbon extraction) well sites will claim only about 2% of the field's area, and this claim can be further minimized with multiple wells drilled directionally from a single pad. Many gathering pipelines transporting hot water and steam are mounted on high supports and do not preclude grazing (or even crop cultivation) and the movement of wild animals, and the plants themselves (including generator halls, cooling towers, auxiliary buildings, and switchyards) are fairly compact: the holding ponds for temporary water discharges during drilling and field stimulation may be the largest land claim component. But pipelines and roads may cause some deforestation, and they obviously contribute to fragmentation of habitat or (in some locations) to increased erosion and a higher likelihood of landslides.

The Geothermal Energy Association put the aggregate claims for the thirty-year operation of a typical geothermal plant at 404 m^2/GWh, that is, 283 W/m^2 (GEA 2013). DiPippo (1991) put the land requirements of a 110-MW geothermal flash plant (excluding wells) at nearly 14 ha, or almost 800 W/m^2 in terms of installed capacity; for a smaller (20-MW) binary plant (also excluding wells), his rate was about 700 W_i/m^2, and the power density for a 56-MW flash plant, including all other infrastructures (wells, pipes, roads), was about 135 W_i/m^2. Finally, McDonald and co-workers (2009) put the highest power density (most compact projects) of geothermal power plants (based on American data) at 113 W_i/m^2 and the least compact at just 8 W_i/m^2.

As with other kinds of electricity-generating stations, the land claims of geothermal plants are readily assessed by accessing satellite images; in addition, we have detailed land-use studies for some major geothermal fields. California's Imperial Valley has some of the most compact facilities located amid crop fields: a 40-MW Heber binary plant (using hot water to heat another working medium, pressurized liquid with lower boiling point to run the turbine) claims 12.15 ha (for a power density of about 330 W_i/m^2) and a nearby 47-MW Heber double-flash plant (with hot water vaporized in sequence under low pressure) occupies just 9.5 ha, for a power density of almost 500 W_i/m^2 (Tester et al. 2006).

Koorey and Fernando (2010) examined New Zealand's geothermal projects (14 plants, with a total installed capacity of 890 MW) and found the lowest power density of about 70 W_i/m^2 at Ngawha 2008 and the highest one at Poihipi at 670 W_i/m^2. Wairakei—the country's largest installation, with 165 MW, using flash steam with a binary cycle—rated 133 W_i/m^2; and the national mean was nearly 200 W_i/m^2. All of these rates are for directly affected areas, that is, the land exclusively occupied by the powerhouse, other plant installations, and pipe routes. The addition of affected areas, arbitrarily defined by the authors as all land 100 m out from the development area, lowers Wairakei's power density to just 40 W_i/m^2 and the national mean to 45 W_i/m^2, but this count exaggerates the actual impact, as about 40% of affected areas are used as pasture and 14% are forested, an excellent example of concurrent land use.

Iceland's largest geothermal development, the Hellisheidi combined power plant (303 MW_e) and heating plant (133 MW_t, to be expanded to 400 MW_t), taps into the Hengill volcanic system with wells up to 4 km deep (Gunnlaugsson 2012; fig. 3.8. The generating plant is supplied by 21 wells, each one supporting 7.5 MW_e, and four drill holes are on a platform of just 1,200 m^2. The output density of hot water is thus 25,000 W_t/m^2, but considerable infrastructure is required to turn that flux into electricity. The entire project—including access and service roads, hot water production and freshwater injection wells, hot water, water, and steam pipes, steam separator stations, powerhouses, cooling towers, steam exhaust stacks, water tanks, discharges, and a switchyard system, injection areas, and connection to the power grid—covers 820 ha, and with 2.3 TWh of annual generation, its capacity factor is nearly 87%. When only actual electricity generation is considered, Hellisheidi's power density is just 32 W_e/m^2; adding the available 400 MW_t of heat raises that figure to about 660 MW of useful energy (electricity and heat) and implies an overall power density of about 80 W/m^2.

My final example is the world's largest concentration of geothermal plants, in the Geysers area of the Mayacamas Mountains north of San Francisco, where 22 sites with a total capacity of 1.61 GW tap hot water

Figure 3.8
Hellisheidi geothermal plant in Iceland. UN Photo/Eskinder Debebe.

from the productive area of 7,690 ha (BLM 2010). This implies an average power density of just 21 W_i/m^2, but most of the land is undisturbed mountains, forests, and meadows. The Calpine Corporation operates 15 of the Geysers projects with about 700 MW of installed capacity, and its infrastructure includes nearly 240 km of steam and water injection pipelines and about 270 km of access roads (Calpine Corporation 2013b). Even with generous average claims (3 ha/plant, 5-m ROW for lines and 10-m ROW for roads), the overall claim comes to about 1,250 ha, and the power density rises to 55 W_i/m^2, very similar to that of New Zealand's Wairakei plant.

In the future enhanced geothermal systems—where high-pressure injections of water and chemical into vertical wells (in a process akin to hydraulic fracturing in oil and gas production) create new fractures in hot rocks—will be able to either boost the productivity of some existing fields or allow the production from otherwise uneconomical reservoirs (USDOE 2014). Eventhe lowest power densities, those between 20 and 50 W_i/m^2, are of the same magnitude as those of alpine hydropower stations and an order of magnitude higher than those of the best large wind farms. Of course, these comparisons ignore different qualitative aspects of these three kinds of claims. In addition, and as in the case of underground coal mining (to be noted in the next chapter), there may be also surface impacts following continuous withdrawals of hot water and reservoir recharging. With outflows much larger than recharges, surface subsidence takes place in many geothermal formations: in parts of the Wairakei field it was as much as 45 cm/year (Allis 1990). Subsidence and landslides should not be a problem with EGS operations, but some of them may experience induced seismicity.

While geothermal energy will remain a globally marginal source of electricity—its share was a mere 0.33% of the global supply in 2010 (Bertani 2010)—it can provide significant shares of household, commercial, and industrial heat at local and regional levels. More than sixty countries are now harnessing this resource. The US capacity of direct heat uses more than tripled between 2000 and 2010, from about 3.8 GW_t to 12.6 GW_t; other large decadal increases have been in the Netherlands (from just 308 MW to 1.41 GW), Sweden (from 377 MW to 4.46 GW), and China (from 2.3 GW_t to 8.9 GW_t), and the global capacity increased from about 15.1 GW_t in the year 2000 to 50.6 GW in 2010, while the actual energy use more than doubled, to 438 PJ, implying an average utilization factor of about 28% (IGA 2013b).

Household heat pumps—with relatively small individual capacity (on the order of 10 kW of heat for a typical American house) but with large aggregate numbers owing to their increasing popularity in affluent countries (in the United States, their total has surpassed one million)—are the single largest category of these direct uses (Lund et al. 2003; Navigant Consulting 2009). The space requirements of these installations (wells, piping, sometime also water tanks and heat exchangers) for commercial uses are minimal and easily accommodated within the fenced facilities. A typical American house will need three holes (up to 150 m deep) at least 6 m apart, or at least 100 m^2 with access space, to produce a heat flux of 100 W/m^2. Horizontal closed loops buried at a depth of 1–2 m cover at least 250 m^2 (a square of 15.8 m, easy to accommodate in most US suburban house lots but too large for older, densely spaced urban housing) and yield a heat transfer power density of 40 W/m^2.

4   Fossil Fuels

Fossil fuels are enormously concentrated transformations of biomass, and hence the power densities associated with their extraction are unrivaled by any other form of terrestrial energy. But gathering, preparing, processing, and transporting these energy sources require a variety of permanent and temporary infrastructures that necessarily dilute the power densities of their delivery for final conversion to heat, motion, and electricity.

All fossil fuels are abundant, and although they are obviously finite (for more than a century the rate of their consumption has been surpassing the rate of their formation by many orders of magnitude), there is no imminent global danger of running out of coal or hydrocarbons. Inevitably, the progressive extraction of fossil fuels led some nations to shift from poorer, and more expensively produced, domestic resources to cheaper imports originating in giant surface coal mines and giant oil and gas fields, often on different continents. Many coalfields, some of them centuries old, have been entirely abandoned as underground mining of relatively thin seams became uneconomical, and an increasing share of hydrocarbons is produced from greater depths, in locations more remote from principal markets or in deeper offshore waters. All varieties of fossil fuels have thus became more expensive, but also more widely available and globally traded on massive scales, and their aggregate extraction is still increasing.

The genesis of fossil fuels explains the often extraordinarily high densities with which these resources are stored in the uppermost layers of the Earth's crust. Coal seams are lithified and compressed layers of ancient phytomass. As most of the original oxygen and hydrogen in the phytomass were driven away by long spans of pressure and heat processing, generations of those ancient swamp forests were concentrated into layers of carbon. Some have remained remarkably pure, others were later adulterated with incombustible rock, and the younger coals still contain plenty of moisture. Relatively large shares of the carbon that were present in ancient phytomass are preserved in coals, up to 15% for lignites and at least 10% for bituminous coals (Dukes 2003). Actual recovery rates of this ancient carbon depend on the mining techniques (discussed later in this section), but the global mean is around 10%. When inverted, this share means that about 10 units of ancient plant carbon (no less than 5 and as many as 20) yield one unit of carbon in extracted coal.

The genesis of hydrocarbons has been entirely different: mixtures of liquid compounds accompanied by lighter gases originated through bacteriogenesis (anaerobic microbial metabolism) and thermogenesis (heat decomposition) of nonhydrocarbon organic molecules (carbohydrates, dominated by cellulose, proteins, and lipids) produced during the eras of high photosynthetic activity and buried in marine or lacustrine sediments; the combination of these kerogen-forming processes can be dated in some shales as far back as 3.2 billion years ago (Rasmussen et al. 2008). Sedimentary source rocks may contain as much as 10% kerogen by mass, but a 1%–2% share is more common. Their prolonged heat decomposition produced liquid hydrocarbons, most of which were subsequently degraded by slowly acting thermophilic bacteria, which convert lighter fraction to denser oils.

The formation of hydrocarbons preserves much lower shares of the original organic carbon than the genesis of coal, often less than 1% and only exceptionally more than 10% during the formation of oil-bearing sediments. After bacterial and heat processing this figure drops by at least another order of magnitude, and usually only a small fraction of mobile hydrocarbons migrates from the source rock to oil and gas reservoirs, from which anywhere between 25% and 50% of fuels in place could be recovered. Using, again, inverted rates, this means that on the order of 10,000 units of carbon present in the original biomass will be required for every unit of carbon in extracted crude oil, and 12,500 units are needed for every unit of carbon in natural gas (Dukes 2003).

Some natural gases require minimum processing before combustion (stripping H_2O, H_2S, and other trace gases); others contain a relatively high share of natural gas liquids (ethane to pentane), which are removed and sold separately. Crude oils must be refined to produce the most desirable transportation liquids, including gasoline, kerosene, and diesel fuel. Inevitably, refining further lowers the overall carbon recovery factor. Dukes (2003) summed up this low rate of carbon transfer by noting that every liter of gasoline (that is, about 640 g of carbon) requires about 25 t of initially sequestered marine biomass, or at least 12 t of carbon, corresponding to a transfer rate of a mere 0.005% and the requirement of roughly 20,000 units of initial biomass carbon to produce a unit of carbon in gasoline.

But as a result of the long periods of ancient biomass accumulations, even these low transfer rates translate into often enormous energy storage in coal seams and in naturally pressurized oil and gas reservoirs, and huge amounts of oil remain bound in their source rocks, mostly in sands and shales. This means that even though some forms of fossil fuel exploitation are relatively space-intensive—notably the surface mining of coal buried under thick layers of overburden and the pumping of oil from large numbers of low-productivity oil wells that also require many access roads and gathering pipelines—the typical power densities of coal and hydrocarbon extraction are high, and the enterprises producing these fuels claim relatively small amounts of land.

Coal Extraction, Preparation, and Transport

Coals form a rather heterogeneous group of fossil fuels, much more diverse than crude oils and natural gases. Considerable ranges apply to the presence of moisture (from less than 1% to more than 40%), ash (incombustible minerals, from less than 1% to more than 40%), and sulfur (from a trace to more than 5%), and even energy density, one of the two key variables that determine the power density of coal resources, has more than a threefold range. Anthracites (essentially carbon with minimal impurities) have an energy density (all rates are higher heating values) of about 30 GJ/t, high-quality bituminous coals range from 24 to 30 GJ/t, subbituminous coal (commonly used for electricity generation) ranges between 18 and 24 GJ/t, and, owing to their high moisture and ash content, lignites (brown coals) span an even wider range, all the way to less than 10 GJ/t (Smil 2008).

The other key variable in appraising coal's recoverable deposits is the thickness of coal seams, which ranges from less than 30 cm, too thin to be exploited by modern mining methods (but commonly extracted with the help of simple tools by hard and dangerous manual labor in traditional underground mines), to stunning near-surface accumulations of coal, where single seams may, as in Victoria's Latrobe Valley, be more than 150 m thick, and stacked coal seams may add up to 250 m (Australian Government 2012). In contrast, modal values for seams that can be extracted only by underground methods are only between 0.5 and 2 m. For example, the northern Appalachian Pittsburgh coal bed has extensive blocks up to 1.9–2.4 m thick (USGS 2013). The combination of these two variables, energy density and the thickness of coal seams, determines the theoretical extremes of the power densities of coal extraction. A thin, 1-m-thick seam of German *Braunkohle*, low-quality lignite with an energy density of 8.5 GJ/t and a specific density of 1.2 t/m^3, stores just over 10 GJ/m^2, while a 30-m-thick seam of high-quality bituminous coal (energy density of 28 GJ/t, specific density of 1.35 t/m^3) will contain about 1.1 TJ/m^2.

Most coal basins have multiple exploitable seams, and hence even with the fuel of lower quality their deposits will average tens of gigajoules under every square meter, and the richest basins will have stores an order of magnitude higher. US data show that the average for the five richest Appalachian coal beds—Pittsburgh, Upper Freeport, Fire Clay, Pond Creek, and Pocahontas 3—is almost 200 GJ/m^2 (USGS 2013). The latest evaluation of coal resources in the Powder River Basin in Montana and Wyoming—where the Wyodak seam, with an energy density of 20.3 GJ/t, is between 9.1 and 24.4 m thick—showed original stores up to 400 GJ/m^2 (Scott and Luppens 2013).

In Germany's brown coal–mining region, between Cologne and Aachen, the deposit exploited by the Fortuna-Garsdorf mine yielded about 400 GJ/m^2 during the 1980s, while the region's three mines operating in 2013, Inden, Garzweiler, and Hambach, extracted from seams with a combined thickness of up to 45 m and contained about 450 GJ/m^2 (RWE Power 2013). World records belong to the Australian coal deposits: in the Latrobe Valley of Victoria's Gippsland Basin, brown coal seams that are up to 100 m thick store more than 1 TJ/m^2, and at least 50% more when inferred resources are added (Geoscience Australia 2012). And the thickest parts of the Number 3 seam of high-quality bituminous coal (32 m; 24.5 GJ/t) in Queensland's Blair Athol mine, whose production should cease in 2016 (Rio Tinto 2013), also stored at least 1 TJ/m^2.

Underground Mining

The actual power densities of coal production depend not only on the energy stored in seams but also on the modes of extraction. Traditional underground room-and-pillar technique had to leave about half of all coal in place to support the tunnel roofs, but longwall mining has made it possible to nearly double that rate. This technique, pioneered in Europe and adopted in the United States only during the 1970s, uses large shearers (drum-shaped cutting heads) to cut coal from the length of a seam (some longwalls are now 300 m long) and dump it onto conveyors while the miners stay protected under a jack-supported steel roof that advances along a seam as the cutting progresses, and the roof behind it is left to cave in (Osborne 2013). Longwall mining can recover more than 90% of coal in place as long as the seams are reasonably level or slightly inclined. This technique boosted productivity to such an extent that the two largest underground US coal mines, Bailey and Enlow Fork, in Pennsylvania, now produce close to 10 Mt/year (USEIA 2011a).

Coal mining land claims fall into two obvious categories. Land occupied by permanent structures includes all buildings (with stationary operating equipment, maintenance shops, and parts stores), parking lots, facilities to process (wash, sort, crush) raw coal before marketing, and load-out arrangements to transport the fuel. The extent of this claim may change with time as a particular operation grows or declines, and in the case of underground mining it can be fairly limited. The aboveground structures of underground mines include buildings to operate hoisting and ventilation machinery, offices, machine shops, and storage spaces. For a mine producing high-quality coal that need only be crushed to uniform sizes before marketing, the only other permanent structures could be a crusher and a silo to store coal and loading facilities for trucks or railroad cars; all of these could occupy only about 1 ha for a mine producing 1 Mt/year.

But most coals need preparation (washing, rock removal, and crushing to specific sizes), and most underground mines have on-site facilities to perform those tasks (Arnold, Klima, and Bethel 2007; Leonard and Leonard 1991). The resulting waste makes up the second, incremental category of land claims as more space is taken every year by its disposal (in the past, it was often stored in tall conical heaps) and by often extensive tailing ponds. Surface mining may have the same kind of need to store waste products, but most of its incremental space demand results from the need to remove and reposition often large volumes of the overburden (sands, clays, and rocks) that covers coal seams.

Even so, large underground coal mines make relatively limited land claims, an obvious consequence of concentrating the output from a large area of underground corridors into a fairly small area of the mine's aboveground operating and fuel-processing facilities. The total for the Bailey Mine in Waynesburg, southwestern Pennsylvania, is about 60 ha, and with an annual output of 9.8 Mt (at 30 GJ/t), that translates into a very high power density of about 15,000 W/m^2. Similarly, even after counting all land claimed by past storage of mining waste and tailings, the Enlow Fork Mine, the second largest underground operation in the United States, also in Pennsylvania, occupies roughly the same area, and its extraction power density is about 14,000 W/m^2. Rates for smaller mines with poorer coal in thinner seams are an order of magnitude lower, but commonly more than 2,000 W/m^2.

But in many relatively shallow mines the tunneling has additional surface impacts far away from a mine shaft as a result of ground subsidence. For example, by the end of 2009 the underground operations of the just mentioned Bailey and Enlow Mines in Pennsylvania extended over respectively 12,600 ha and 14,000 ha, with further areas intended for expanded longwall mining (Schmid & Company 2010). Surface subduction caused by underground mining is sometimes a continuous slow process that damages streets and buildings; in other cases it results in a sudden catastrophic collapse. The impact on local water resources is another surficial effect of underground mining that is of obvious concern to communities relying on those water supplies.

Surface Mines

The removal of the overburden in surface (open-cast) mines has become an increasingly ambitious enterprise. Only in exceptional cases are thick coal seams hidden beneath a thin layer of soil and rocks: in Australia's Latrobe Valley, thick brown coal seams lie under only 10–20 m of easily removed sandy overburden (Minerals Council of Australia 2011). Technical advances have made it possible to exploit reserves whose overburden to coal seam ratio is higher than 3:1, in a few cases even surpassing 6:1, and the deepest mines (such as the Rhineland's Hambach) now extend to as much as 370 m below the surface.

Topsoil from fields is put aside for future reuse or is used in ongoing recultivation projects on the site, and in Europe, entire villages or small towns may have to be relocated or their inhabitants resettled elsewhere (Michel 2005). The overburden (usually sedimentary clays and rocks) is removed by some of world's largest electric machines: stripping is done either by giant bucket excavators (common in German mines and deployed to create terraced cuts) or by walking draglines (used in the United States). The German practice is to deposit the overburden by spreaders behind the cut or carry it farther away by conveyor belts and use it to backfill older mined-out sites or to create artificial tabletop accumulations. In the United States, surface mining in the West, including the world's largest open-cast mines in Wyoming, is done by terraced trenches or pits in ways similar to German operations.

In Appalachia, miners follow the contours of hills with outcropping seams, or they proceed in a much more destructive way: since the 1980s, mountaintop removal has leveled more than 500 peaks, mostly in Kentucky and West Virginia (Copeland 2014; McQuaid 2009; Perks 2009). Summits or ridges above coal seams are removed by blasting away layers of rock often more than 150 m thick. Laws require returning this overburden to the mined areas, but the spoils are commonly dumped into adjacent valleys, creating massive fills that bury streams and may be more than 300 m wide and more than 1.5 km long, with a volume of more than 200 Mm^3. The extensive land claims of mountaintop mining are also illustrated by averages of land occupied per mine in West Virginia: for underground mines it is just over 16 ha, whereas for mountaintop operations it is about 140 ha.

Coal is usually mined by giant electric shovels (with a bucket able to scoop more than 100 t), and its recovery rates are high: in the world's largest mines, in Wyoming, they average 91% (WSGS 2013). Cut coal is either transported from the coal face by conveyor belts (in Europe) or it is hauled away by huge off-road Liebherr, Komatsu, or Caterpillar trucks capable of carrying loads of more than 300 t. In some mines coal is moved by conveyor belts directly from the mining face to a nearby mine-mouth electricity-generating plant, but in most cases more land is claimed by loading facilities to transport the fuel by rail: for long-distance deliveries it is rapidly loaded into railway cars coupled in unit trains of up to 150 cars (Khaira 2009).

The trains' total loads are usually in excess of 10,000 t, and loading is done in less than two hours. In large open-cast Wyoming mines these loading arrangements consist of a railway loop, tall coal silos, and automatic loading dispensers that operate round the clock. In the Black Thunder Mine the loading facility (including the railroad loop) occupies less than 70 ha, in the Cordero Mine less than 20 ha. Unit trains run constantly between a mine and their destinations, large thermal electricity-generating plants that might be hundreds, even thousands of kilometers away, and coastal loading terminals, where coal is transferred to large bulk carriers for intercontinental shipping (mostly from Australia, Indonesia, South Africa). Some long-distance transport is also done very inefficiently by trucks (a common practice in China, as railroads are already overburdened by coal shipping) and very efficiently by river barges (in the United States on the Mississippi).

Some open-cast coal mines publish their cumulative and annual land claims, and satellite imagery makes it easy to assess not only the extent of the areas directly affected by ongoing open-cast extraction but usually also of the previously exploited areas that have been simply abandoned, replanted with grasses and trees, or converted into water reservoirs. Calculations based on the company's data (RWE Power 2013) show the following annual power densities of lignite extraction in the three Rheinland open-cast mines: Inden, 2,400 W/m^2; Garzweiler, about 2,800 W/m^2; and Hambach, approximately 5,000 W/m^2 (fig. 4.1).

For the world's two largest open-cast mines, the North Antelope Rochelle Mine, a complex of two mines (operated by Peabody Powder River Mining, near Wright, Wyoming; fig. 4.2) and the Black Thunder complex (Arch Coal Company, operating six pits), recent power densities of coal extraction have been almost 12,000 W/m^2 (Arch Coal 2013; Wyoming State Geological Society 2013). In 2010 and 2011, Queensland's Blair Athol Mine extracted coal with densities around 2,400 W/m^2 if only incremental land claims were counted (Rio Tinto 2013). China, the world's largest coal producer, extracts most of its coal from underground mines, but its largest open-cast operation, Heidaigou Mine, south of Junggar in Nei Monggol, produces subbituminous coal (at an energy density of 16 GJ/t) from a site that now covers about 15 km^2, with a power density of less than 1,000 W/m^2 (Google Earth 2014; Zhang and Cotterill 2008).

Figure 4.1
Garzweiler lignite mine, Germany. © FEDERICO GAMBARINI/epa/Corbis.

Mountaintop removal is unquestionably the most space-demanding method of coal extraction (also causing great harm to streams, water quality, and biodiversity), as many of these operations cover more than 500 ha and as the largest mountaintop mines can claim more than 2,500 ha and result in dumping some 750 Mm^3 of spoils. Extraction in some of these mines—where massive volumes of rock are blasted away to get at some thin seams—proceeds with power densities lower than 200 W/m^2, and even less than 50 W/m^2, the same order of magnitude as some PV conversions. I have chosen extreme cases for these calculations in order to establish the full range of power densities of coal production, and there is a helpful way to verify the results by using aggregate US data on land disturbed by coal mining.

According to the annual reports of the US Office of Surface Mining and Reclamation, between 1996 and 2009 surface mining in Wyoming disturbed about 25,700 ha, for a cumulative output of 4.37 Gt (SourceWatch 2011). Converting the output with an average energy density of 20.3 GJ/t yields a high power density of almost exactly 11,000 W/m^2 for Wyoming

Figure 4.2
North Antelope Rochelle Mine Complex, Wyoming. NASA Earth Observatory.

coal extraction. In contrast, in 2009 the power density for surface coal mining in Tennessee was only about 350 W/m^2, a very low rate caused by the state's extensive mountaintop extraction. Land claims in other states fall between these two extremes, and nationwide totals of 624,400 ha and 8.69 Gt for the years 1996–2009 translate to just over 1,000 W/m^2.

Those open-cast operations that follow the strictest operating rules cause only temporary disruption. Once the mining stops, the land can be contoured to approximate its natural state and replanted with grasses, shrubs, and trees or converted to water surfaces within three to five years after the mining ends. On the other hand, every coal-mining country has large areas of old open excavations, overburden heaps, rock waste from underground mining, and mine tailings that have never been reclaimed. Since 1977, when the US Congress passed the Surface Mining Control and Reclamation Act, the program has reclaimed nearly 100,000 ha of land affected by coal mining, but that still leaves almost 5,200 coal-related abandoned mine sites to be fully reclaimed (Abandoned Mine Lands Portal 2013).

Coal Transportation

Adding land needed for coal transportation outside mine loading facilities and unloading structures at power plants of ports could be done easily only in the case of a dedicated line used for no other shipments—but then it might not turn out to be a significant addition. For example, a double-track line used only to move coal between an open-cast mine (extracting 10 Mt/year with a power density of 2,500 W/m^2) and a 4-GW_e power plant 500 km away would occupy (even under the liberal assumption of a 20-m right-of-way for two tracks) 1,000 ha—but during the 30 years of extraction the mine would disturb more than 11,000 ha.

In reality, a large mine usually supplies many customers, and unit trains travel over many lines and share them with other traffic (Kaplan 2007). For example, coal from the Black Thunder complex is transported via Burlington Northern Santa Fe and Union Pacific railroads to some 115 power plants in more than 20 states, as well as to Europe (BNSF 2013). While it might be possible to calculate the share of the ROWs attributable to coal that has originated from that mine since it began coal extraction (on the basis of a detailed breakdown of annually carried cargo), it would, when prorated over many decades, represent a negligible addition to incremental space claims made by that large open-cast mine. Similarly, land occupied by large railway terminals where coal is received and loaded for overseas export is a small fraction of the land claimed by mining. The Lamberts Point Coal Terminal in Norfolk, Virginia, for example, covers more than 150 ha, but because it can handle up to 44 Mt of exports a year (Dinville 2013), its throughput power density prorates to at least 24,000 W/m^2, a very high rate, adding a negligible amount to the overall land claim.

Perhaps the most notable conclusion of this survey of power densities of modern coal extraction is their wide range. The most productive underground mines using longwalls produce the fuel with power densities in excess of 10,000 W/m^2, while many underground mines exploiting thinner seams of poor-quality coal and storing incombustible waste and tailings produced by the requisite coal cleaning have power densities well below 1,000 W/m^2. And the differences are even greater for surface mines: some mountaintop removals in Appalachia rate well below 100 W/m^2 (fig. 4.3), while extraction in some of the world's largest mines exploiting the world's richest coal seams in Wyoming's Powder River Basin or Australia's Latrobe

Figure 4.3
Mountaintop removal coal mining, in West Virginia. © Daniel Shea/Galeries/Corbis.

Valley exceed 10,000 or even 15,000 W/m^2. The power densities of coal extraction thus span two orders of magnitude, being nearly as low as those of PV-based electricity generation and as high as those of rich hydrocarbon fields.

Crude Oils, Refining, and Long-Distance Deliveries

Crude oils are much more energy dense than coals (their energy densities cluster tightly around the mean of 42 GJ/t) and their share of incombustible constituents is negligible, but their sulfur content is often high (more than 2%) and serves to divide the liquids into sweet (low-sulfur) and sour (high-sulfur) streams (Smil 2008). Another key classification is according to specific density, with light oils containing a higher share of lighter fractions and heavy oils requiring expensive catalytic cracking to produce higher shares of the most valuable transportation fuels, gasoline and kerosene. The worldwide supply of crude oil has been facilitated by relatively inexpensive ways of long-distance delivery, by pipelines on land and by tankers for the intercontinental trade.

Prorating the best reserve estimates per unit of surface area that corresponds to the underground extent of the reservoirs results in power densities that are very similar to those of coal deposits. For many smaller fields the rates are less than 1 GJ/m^2, and even some giant oil fields (the designation applies to any field that contains 500 million barrels, or 59,000,000 m^3, of ultimately recoverable oil) store less than 10 GJ/m^2. Only the richest fields store 10^1–10^2 GJ/m^2. As many fields also contain substantial volumes of associated natural gas, their original storage should be reported for the combined content of the two hydrocarbon fuels. The next paragraph lists a few well-known examples of original storages in ascending order of energy density (Li 2011; Nehring 1978; Robelius 2007).

The Algerian Hassi Messaoud oil field, known for its exceptionally light and sweet crude, rates 35 GJ/m^2. The Shayba oil field, a supergiant in the desolate sands of Rub' al-Khali/Empty Quarter desert, Saudi Arabia, stores about 130 GJ/m^2. Samotlor, Russia's largest oil field and number six in the global ranking, which covers 1,752 km^2 in the Tyumen region of western Siberia, had original stores of about 180 GJ/m^2. The al-Ghawar oil field, the world's richest oil field and a true supergiant that extends over 220,000 ha in Saudi Arabia's Eastern Province, originally contained about 260 GJ/m^2 (220 GJ/m^2 of crude oil and 40 GJ/m^2 of natural gas). Alaska's Prudhoe Bay (North Slope) oil field, the largest (85,417 ha) and richest hydrocarbon field in North America, had originally in place about 145 EJ of crude oil and 44 EJ of natural gas (BP 2006), for an aggregate energy storage density of about 220 GJ/m^2. And the Greater al-Burqan, Kuwait's largest field and the world's largest petroliferous sandstone formation, just south of Kuwait City, which covers 780 km^2 and had about 440 EJ of oil and 65 EJ of natural gas originally in place, stored about 560 GJ/m^2.

Most of the oil in the Earth's crust is not in liquid form but is interspersed in sands or shales, and the storage density of these nonconventional oil resources rivals that of the richest classical fields. This is hardly surprising, as the shares of oil in these rocks are low (commonly less than 10%) but the oil-bearing strata can be very thick. The world's largest concentration of oil interspersed in shales is in the Green River formation, which underlies parts of Colorado, Utah, and Wyoming; the oil-bearing layers there are up to 150 m thick. The Piceance Basin in Colorado holds the world's largest deposit of shale oil: there are roughly 210 Gt of oil locked in the rock under its roughly 18,200 km^2 (Johnson et al. 2010), which prorates to nearly 500 GJ/m^2 of energy originally in place. That is more than twice the density in the world's richest field producing liquid oil, but the technical challenges and costs (and ultimately the low net energy returns) of extracting oil from Colorado shale are far greater than those associated with getting oil from giant oil fields where the oil initially flows under natural pressure.

Power Densities of Oil Production

Extraction densities depend on the richness of exploited reservoirs, methods of oil recovery (free-flowing wells, artificial lift aided by water flooding or injection of gases), the density of wells, and their productivity. Crowded spacing of oil wells was common in the early decades of oil era (fig. 4.4); now the spacing is carefully planned to optimize the output, which similarly is deliberately controlled either to extend the production span of an oil field or to match the available pipeline capacity. The temporary oil yield

Figure 4.4
High density of producing oil wells in the Spindeltop field near Beaumont, Texas. ©
Bettmann/CORBIS.

of some the world's most famous oil well gushers was astounding (SJVG 2010). The earliest gushers were in the Baku oil fields: Bibi Eibat, drilled on September 27, 1886, flowed at 84,000 barrels per day (bpd), which amounts to (at 5.8 GJ/barrel) to a daily rate of about 5.6 GW .

Starting on October 15, 1927, the Baba Gurgur oil field near Kirkuk, Iraq, spewed 95,000 bpd for eight and one-half days (6.4 GW). The famous Spindletop, near Beaumont, Texas, produced 100,000 bpd on January 10, 1901 (6.7 GW), while the recorded maximum is for Cerro Azul No. 4 well near Tampico, Veracruz, Mexico, which blew on February 16, 1916; when it was capped, three days later, its February 19 flow was 260,858 bpd, or 17.5 GW. For comparison, BP's Macondo well, which spilled nearly 700,000 t of oil into the waters of the Gulf of Mexico between April 20 and July 15, 2010, had a maximum flow of about 62,000 bpd, or 4.2 GW.

And all of these gigawatt-sized blow-outs spewed from pipes whose area was no more than 0.02 m^2! Even if we were to prorate such flows across a large drill site of 2 ha (200 × 100 m), we would get power densities on the order of 10^5 W/m^2 (300,000 W/m^2 for a 6-GW gusher). No other power densities of energy extraction are even remotely close to these extraordinary releases of accumulated stores of high-energy-density fossil fuels. But all of them are necessarily short-lived, lasting only for hours or days, and the output of every naturally pressurized well will decline, usually following a fairly predictable course. Secondary recovery methods will help, but they will claim more land for new injection wells, as well as for the delivery of water or gases.

Perhaps the most representative examination of land claimed by conventional crude oil production comes from the data set of 301 California oil fields covering some 3,000 km^2 and containing at least 58,000 production wells, 22,000 shut-in wells, and 25,000 injection wells, at an average density of 31 wells/km^2 (Yeh et al. 2010). Calculations of land disturbed per well were based on satellite image analysis of three fields and resulted in the range of 0.33–1.8 ha/well, with an average of 1.1 ha/well, and that rate included not only the cleared or occupied area surrounding a well but also all access roads and other facilities found in each image. The authors used the identical approach to assess land claims of Alberta's conventional oil fields, whose much lower density of only 0.3–2.5 wells/km^2 translated into larger land claims, ranging from 1.6 to 7.1 ha/well and an average of 3.3 ha/well (Yeh et al. 2010).

Satellite images of sufficiently high resolution (showing objects smaller than 5 m) indicate that in the world's most productive fields, the extent of land disturbance per well is closer to the just noted Californian mean rather than to the Albertan average. Saudis have not been forthcoming with detailed information about the development and production of the world's largest supergiant oil field, but we know that al-Ghawar has about 3,400 wells that dot its elongated (along the NNE-SSW axis) shape in the eastern Arabian desert between Fazran and Haradh (Afifi 2004), and that these wells typically are rectangular areas of 100–150 m (1.5 ha) graded in sand (Google Earth 2014). The field also has many water injection wells—with water brought by pipelines from the Persian Gulf—as water flooding has become increasingly necessary to keep up its productivity. Samotlor's wells, dotting the western Siberian forest and wetlands, claim mostly between 0.5 and 2 ha.

A detailed study of land disturbance associated with oil and gas development in the Upper Colorado Basin found that the lifetime average pad size in Utah, Colorado, and Wyoming was 1.04 ha (Buto, Kenney, and Gerner 2010). But considerably more land could be disturbed if there is a need to build new long access roads and to clear land for larger storage yards. As for the oil well densities, their maxima are prescribed in many jurisdictions. For example, in the East Texas field, rules require a minimum distance of 200 m between adjacent wells, which translates to a maximum of 25 wells/km^2; the West Texas field also has more than 20 wells/km^2. Oklahoma assigns 16 ha as the area drained by shallower wells and 32 ha for deeper wells (implying densities of respectively 6.25 and 3.12 wells/km^2), while some smaller fields in the United States have more than 50, others less than 5 wells/km^2.

Lack of accurate information makes it impossible to offer reliable calculations of peak and short-term extraction densities for the world's largest supergiant oil fields. Assuming that al-Ghawar's output peaked at 250 Mt/year, its maximum extraction power density would have been only on the order of 150 W/m^2 when that output is prorated over the entire footprint (2,200 km^2) of the reservoir, but it rises to about 5,000 W/m^2 when only areas disturbed by wells and roads are considered. Most of al-Burqan's 350 wells (famously set on fire by Saddam Hussein's retreating army in 1991) have areas of about 1 ha, but the total reservoir area is given either as 500 km^2 or as much as 820 km^2 (with two smaller associated formations included), and the field's peak output was either 120 or

140 Mt/year in 1971 or 1972 (Sorkhabi 2012). Consequently, the field's maximum operating power density was as much as 370 W/m^2 (prorated over the reservoir's area), and more than 20,000 W/m^2 when only areas disturbed by wells and roads are considered.

But in order to convert land claims to truly representative power densities of oil extraction we need to know the long-term productivities of the oil wells. All natural well flows decline with time, and hence the power densities for the early years of extraction will be substantially higher than in later years or decades. Again, no accurate rates can be offered for the largest Middle Eastern reservoirs because the total mass of their ultimately recoverable oil and the complete duration of their exploitation are uncertain. For example, the total for al-Burqan is as low as 6 Gt and as high as 10.5 Gt, and the field is still producing at about half its peak rate nearly 70 years after the beginning of its exploitation.

The best way to gauge long-term power densities is to look at the history of those American oil fields that have been exploited for many generations. For example, the power density of production from the California oil wells studied by Yeh and co-workers (2010) averaged about 2,500 W/m^2 for the cumulative 1919–2005 output but only about 1,700 W/m^2 for the 2005 extraction, while the historical (1958–2007) power density for Alberta oil wells was roughly 1,100 W/m^2, with the marginal rate for 2007 declining to about 640 W/m^2. Data for the evolution of the West Texas Wasson oil field near the New Mexico border (Smith 2013) illustrate this process of gradual decline in greater detail.

The field's first well, drilled in 1936, had a short-term-production power density of about 1,500 W/m^2. By 1938, 201 flowing wells (assuming 1 ha/well) averaged about 250 W/m^2, and the field's peak output in 1948 prorated, with 1,588 wells, to about 320 W/m^2. The field is clearly outlined by its fairly regularly spaced wells (fig. 4.5): many are in an almost perfectly regular square grid pattern and occupy usually no more than 2,500 m^2 (0.25 ha), which means that by 1992, the field's 2,242 wells (and their service roads) claimed only some 900 ha, with the addition of several associated gathering and processing facilities raising the total to about 1,000 ha. The power density of the field's cumulative output over 56 years of extraction would have been about 600 W/m^2.

US data illustrate changing productivities on a national scale. The average productivity of US oil wells rose from 13 bpd in 1955 to a peak of 18.5 bpd in 1972, and since that time has declined to 10.8 bpd in 2000 and 10.6

Figure 4.5
West Texas Wasson oil field. NASA Earth Observatory.

bpd in 2011 (USEIA 2012). The decrease would have been much higher without the extensive use of secondary recovery, now a standard practice in all aging oil fields that relies on injections of water or gases to boost oil flow (SPE 1999). In contrast, directional and horizontal drilling make it possible to reach larger volumes of oil-bearing layers from a single well and thus spare the land and boost the average power densities of extraction. But these innovations, even when combined with new, highly productive discoveries, could not prevent the global decline of productivity, and hence worldwide reductions in the power densities of oil extraction.

Oil & Gas Journal has been publishing annual global reviews listing the numbers of producing wells and average production rates for oil-producing nations as well as for major oil fields. In 1972, the last year of inexpensive oil (in the fall of 1973 came the first of OPEC's large price increases, which quintupled the cost of a barrel), Saudi Arabia had only 627 wells, and their output, even when assuming land claim of 2 ha/well, prorated to about 40,000 W/m^2, while average power densities for the entire Middle Eastern oil extraction were nearly 25,000 W/m^2, and the worldwide mean (excluding the USSR) was around 500 W/m^2. This low global figure was largely the result of the low productivity of many thousands of old and older US wells: even if a land claim of just 1 ha/well is assumed, the US mean in 2012 was only 125 W/m^2. Clearly, J. E. Mielke's contemporaneous estimates of land claims by the US onshore oil extraction—3.03–6.9 acre-year/10^{12} Btu, that is, 1.2–2.8 ha/PJ or 1,100–2,600 W/m^2 (Mielke 1977)—referred to well sites occupied by highly productive new wells.

Oil & Gas Journal data for the year 2012 show that the power densities of oil extraction had fallen everywhere, with the means (even when assuming just 1 ha/well) at about 23,000 W/m^2 for Saudi Arabia, less than 9,000 W/m^2 for the Middle East, and about 100 W/m^2 for the United States, with the global mean, including all of the world's countries, at about 650 W/m^2 (*Oil & Gas Journal* 2012). The power densities of oil extraction thus range between 10^2 W/m^2 for old, or older, oil provinces outside the Middle East (some of them, including Azerbaijan, California, and parts of Texas, have been exploited for more than a century) to 10^4 W/m^2 for the most productive fields on the Arabian Peninsula. The decline of average power densities has been unmistakable, and although this trend has been slowed down by secondary recovery, it clearly illustrates the maturity, even advanced age, of many of the world's major oil fields.

Unconventional Oil Production

Declining oil well productivities in conventional fields do not signify any imminent exhaustion of liquid hydrocarbons because vast volumes of oil are locked in solid rocks, sands, shales, and tars. Some of these vast kerogen resources can be now tapped by modern extraction methods. Recent increases in US crude oil production have been mostly attributable to crude oil from shales extracted by a combination of horizontal drilling and hydraulic fracturing (American Petroleum Institute 2009; USEIA 2011c). The Bakken shale, part of the Williston Basin, located mostly in North Dakota and Saskatchewan, has been the fastest-developing new oil development in the United States, with more than 5,000 new wells drilled in the five years starting in 2009 (Patterson 2013). New Bakken well sites average 2 ha, and subsequent reclamation reduces that to about 0.8 ha for production that draws on a subsurface area of 512 ha of oil-bearing shales; in contrast, conventional vertical drilling would claim 4–20 times the surface area.

But while properly managed conventional wells can maintain a fairly steady or a slowly declining output for many years, production from fractured horizontal wells is characterized by rapid, hyperbolic declines. For example, during their first year of production the wells in North Dakota's Bakken oil field could produce as much as 2,000 bpd, which was followed by 65%–80% declines in subsequent years (Sandrea 2012). A typical Bakken well yields 900 bpd during the first year, less than half that much in the second year, only 65 bpd during the fifth year, and 40 bpd during the tenth year (Likvern 2013). As a result (when assuming an average well area of 1.5 ha), the cumulative power densities of its oil extraction would be about 4,000 W/m^2 in the first year, roughly 1,600 W/m^2 for the first five years, about 900 W/m^2 for the first decade of its output, and less than 400 W/m^2 for three decades, although many wells will not be in operation that long. The rates would be significantly reduced if new access roads, needed not only to bring in the drilling equipment but also for regular deliveries of fracking liquids, were taken into account.

There is no commercial recovery of oil from the American Green River shales, but Alberta's oil sands already provide nearly 60% of Canada's oil extraction. The three principal formations—Athabasca-Wabiskaw, Cold Lake, and Peace River—cover a total of about 140,000 km^2, with 10.6 ZJ (1.75 trillion barrels) of oil in place (Hein 2013); this translates to a storage energy density of about 75 GJ/m^2. There are two important and distinct ways of exploiting oil sands: surface mining and in situ recovery (fig. 4.6). Surface mining entails the removal of overburden (peat, clay, sand), the excavation of relatively oil-rich sand strata, the transportation of these minerals in giant trucks for oil extraction (using hot water and NaOH), and the ensuing creation of large tailing ponds that now cover 176 km^2 and contain mixtures of water, sand, clay, and residual oil left over after processing (CAPP 2013). Separated bitumen (with densities in excess of 1 t/m^3, compared to less than 0.95 t/m^3 for conventionally produced oil) is first sent for upgrading and then to a refinery (Gray 2001).

Only a tiny share of Alberta oil locked in sands can be extracted by surface mining, and it is expected that eventually some 98% of aggregate production will come from in situ recovery. The first process of this kind was cyclic steam stimulation (CSS), with periods of injecting hot pressurized steam (300°C, 11 MPa) into well bores followed by periods of months to three years of soaking to loosen the bitumen and then pump out the

Figure 4.6
Alberta oil sands. Top: Surface mining of bitumen. © Ashley Cooper/Corbis. Bottom: In situ recovery. © Brett Gundlock/Corbis.

bitumen-water mixture from the wells used for steam injection, with recovery rates of 25%–35%. The most rewarding in situ recovery technique is steam-assisted gravity drainage (SAGD), whereby the oil in place is softened by steam injected into a horizontal well (500–800 m long) and drains through slots to a gathering well placed 5 m below the steam conduit. This method can recover up to 60% of oil in place, but it is both energy- and water-intensive; after separation, water is reused and oil is piped for upgrading.

The power densities for surface mining (with annual energy yields ranging between 0.61 and 1.2 PJ/ha) range from 1,900 to 3,700 W/m^2 (mean, about 2,900 W/m^2). For in situ recovery (with annual yields of 2.2–5.2 PJ/ha) they are substantially higher, between 7,000 and 16,000 W/m^2, with a mean of about 10,500 W/m^2 (Yeh et al. 2010). But all of these density calculations exclude land claimed by natural gas production that is needed to heat water for the extraction of bitumen from excavated sands, to produce steam for in situ recovery, and to provide energy for the upgrading of bitumen. These three requirements average respectively 70, 220, and 50 m^3 of natural gas per cubic meter of upgraded oil, and the inclusion of all of these upstream land claims lowers the typical power density of oil produced by surface mining to about 2,300 W/m^2, and that from in situ extraction and upgrading to less than 3,200 W/m^2 (Jordaan, Keith, and Stelfox 2009; Yeh et al. 2010).

Pipelines and Refineries

Gathering pipelines take oil from individual wells to field storage tanks or to processing facilities. They are relatively short, with small diameters (typically just 5–20 cm); their limited throughputs restrict their power densities; and their ROWs are minimal in compact and highly productive fields, whereas in old fields with a large number of stripper wells (marginal wells approaching the end of their extractive span) their network may be relatively extensive. Canada has about 250,000 km of such lines, roughly 3,800 km for every million tonnes of conventional oil production (CEPA 2013). In contrast, the United States has no more than 65,000 km of gathering pipelines, located mostly in Texas, Oklahoma, Louisiana, and Wyoming (Pipeline 101 2013).

Crude oil is often processed before it enters a pipeline. Processing separates oil from natural gas, and in the fields using water flooding the two liquids must be separated to prevent pipeline corrosion. Some crudes also require desalting and at least a partial removal of H_2S (sweetening) and stabilization before sent into a pipeline. Many fields also have on-site storage tanks of limited capacities. Crude oil is transported by every kind of commercial carrier (with the obvious exception of aircraft): it is often trucked or transported by rail cars and loaded on barges for river transport, but the two leading means of its long-distance delivery are pipelines on land (as well as from offshore fields) and large tankers on the oceans.

Pipelines are not only the least expensive mode of transporting oil on land but also the safest. The United States pioneered their use starting in the 1870s but the longest post–World War II American lines, from the Gulf of Mexico to the East Coast, were eventually eclipsed by pipelines carrying oil from western Siberia's supergiant Samotlor oil field to Western Europe, a distance of about 4,500 km. The United States has an extensive network of both crude oil pipelines carrying oil from major oil basins (with diameters of up to 1.2 m) and from tanker ports and product pipelines moving refined fuels (above all gasoline and kerosene, with diameters of up to 1.05 m) to major consumption centers. Pipeline construction claims corridors 15–30 m wide (needed for trenching, pipe delivery, and the operation of pipe-laying machinery) and ROW strips for operation; on US federal lands they are normally up to 15 m for buried pipelines and 19.2 m for elevated pipelines, but the extremes can range from only about 10 m to more than 30 m.

These strips are kept clear of major vegetation and any major obstructions to ensure access for monitoring, maintenance, and effecting needed repairs. The danger of accidental encroachment at major lines can be prevented, or at least minimized, by aerial or satellite monitoring, by burying fiber cable above the pipelines, or by placing infrasonic seismic sensors (Chastain 2009). Pumping stations, placed at roughly 100-km intervals, need 10–20 ha. The latest statistics show about 80,000 km of crude oil lines and 140,000 km of product lines, and a total transported mass of roughly 1.675 Gt (BTS 2013). Assuming (perhaps a bit conservatively) an average ROW of 15 m, those figures imply an average nationwide throughput power density of 675 W/m^2.

America's longest pipeline is the Trans-Alaska Pipeline System, owned and operated by the Alyeska Pipeline Service Company and linking the North Slope with Anchorage. The line is 1,288 km long (with about 700 km

Figure 4.7
Trans-Alaska Pipeline. © Robert Harding World Imagery/Corbis.

built aboveground through the permafrost territory), and its ROW is about
20 m. Its peak throughput, in 1988, was just over 100 Mt of crude oil, but
by 2012 the rate was down to just over 30 Mt/year (Alyeska Pipeline 2013;
fig. 4.7). These performances translate to throughput power densities of
about 5,300 W/m^2 and 1,500 W/m^2, respectively. Many oil pipelines of
moderate to low capacity will have throughput power densities of less than
500 W/m^2.

Tanker loading facilities make rather limited onshore claims, mainly for
necessary storage tanks; the transfer operations are often located entirely
offshore in order to accommodate larger ships with a deeper draft: the most
famous examples are the Saudi Ras Tanura and the Iraqi Faw facilities. As
a result, the throughput power densities of the largest crude oil–loading
facilities are in excess of 10^5 W/m^2. Refineries are also inherently high-
throughput facilities, but those with large capacities claim substantial
blocks of usually coastal land. Their densely packed assemblies of columns,
piping, and arrays of storage tanks are designed to process annually
10^4–10^7 t of crude oil in a single facility that has often seen gradual expansion into the surrounding areas.

Figure 4.8
ExxonMobil giant Baytown refinery, Texas. © Smiley N. Pool/Dallas Morning News/
Corbis.

The world's largest refineries process in excess of 500,000 bpd, that is, at
least 25 Mt of crude oil a year with an annual throughput of 33.3 GW. These
complexes are readily identifiable on satellite images, and the only uncer-
tainty in calculating their throughput power densities concerns whether to
include the often extensive areas within the facilities that have been abandoned or that are held in reserve for future expansion. Centro Refinador
Paraguaná in Falcón, Venezuela, has the world's largest capacity (940,000
bpd), but in 2013 it was actually processing only 588,000 bpd, and it occupies about 500 ha; that translates to an average throughput power density
of almost 8,000 W/m^2.

The second largest facility, Ulsan in South Korea, has a capacity of
817,000 bpd and a throughput power density of nearly 5,000 W/m^2. The
ExxonMobil giant refinery in Baytown, Texas, America's largest, covers
9.7 km^2 and has a capacity of 560,640 bpd, and hence a power density of
about 3,900 W/m^2 (ExxonMobil 2013a; fig. 4.8). Exxon's Baton Rouge refinery in Louisiana, the second largest in the United States, occupies 392 ha
along the eastern shore of the Mississippi and 840 ha when its tank farm is

included; with 500,000 bpd, those claims translate respectively to 8,600 and 4,000 W/m^2 (ExxonMobil 2007). The Yanbu' refinery on the Red Sea in Saudi Arabia, which produces fuel for the domestic market, occupies 165 ha and processes 225,000 bpd, resulting in a power density of roughly 9,200 W/m^2 (Saudi Aramco 2013). The largest refinery in the Middle East, the Saudi Ras Tanura, with a capacity of 550,000 bpd and an area of roughly 500 ha, has a power density of nearly 7,500 W/m^2, but small facilities will rate well below 1,000 W/m^2.

Natural Gas, Pipelines, and Liquefied Natural Gas

Natural gas is usually a mixture of the three lightest homologs of the alkane series, methane (CH_4), ethane (C_2H_6), and propane (C_3H_8). American analyses show the following ranges of the three gases: 73%–95% for CH_4, 1.8–5.1% for C_2H_6, and 0.1%–1.5% C_3H_8. Some of the heavier alkanes, mostly butane and pentane, may be also present and are separated out as natural gas liquids before gas enters a pipeline. Raw natural gases have energy densities between 30 and 45 MJ/m^3; pure CH_4 contains 35.5 MJ/m^3, or less than 1/1,000 of crude oil's volumetric energy density. The marketed production of natural gas is appreciably lower than its gross withdrawals, and US data show the reasons for the difference: the extraction loss is about 4%, the removal of nonhydrocarbon gases reduces the volume by about 3%, and less than 1% of the aggregate flow is vented, but nearly 12% is used for field repressurization, leaving the final dry gas production at about 80% of initial withdrawals (USEIA 2013c).

The low energy density of methane limits the total amount of energy stored in gas reservoirs, but the formations with thick gas-bearing strata have storage densities comparable to those of the world's largest oil fields. The South Pars-North Dome field, shared by Iran and Qatar, is the world's largest store of hydrocarbons: on top of its 51 Tm^3 of natural gas (about 35 Tm^3 are considered recoverable) it also contained originally nearly 8 Gm^3 of natural gas condensates (Esrafili-Dizaji et al. 2013). This translates (assuming that energy density of 1 m^3 of condensate equals that of 1 m^3 of crude oil) to about 2.1 ZJ of energy, and, with the field's area of 9,700 km^2, it implies a storage density of about 215 GJ/m^2. Europe's largest onshore natural gas field is the Dutch Groningen in the north near the German border (NAM 2009). The field (discovered in 1959, producing since 1963) had an original gas volume of 2.8 Tm^3, with the 100-m-thick reservoir rock underlying about 900 km^2 of countryside, implying a storage energy density of about 110 GJ/m^2.

The west Siberian Urengoy gas field is the world's second largest supergiant gas field, but its initial content of as much as 8.25 Tm^3 is only about 15%, and its recoverable volume (of about 6.3 Tm^3) is less than 20% of the South Pars-North Dome storage (Grace and Hart 1990). The field underlies about 4,700 km^2 of thick Siberian permafrost that turns into summer swamps and lakes, implying an initial storage density of just over 60 GJ/m^2. Yamburg field, north of the Arctic Circle in Siberia's Tyumen region, comes third: with a recoverable share of 3.9 Tm^3 and the reservoir underlying about 8,500 km^2 of tundra, its initial storage density was roughly 35 GJ/m^2. Hugoton, America's largest natural gas field, is an elongated formation that extends from southwestern Kansas through Oklahoma to Texas; originally it contained 2.3 Tm^3, or about 20% less than Groningen, but its large area of nearly 22,000 km^2 reduces its storage density to less than 4 GJ/m^2 (Dubois 2010).

Natural gas also comes from three other major sources. The most common one is associated gas whose flows accompany the crude oil extraction (wet gas). This gas is dissolved in the crude oil, and after it reaches the surface, the heavier alkanes are separated as natural gas liquids before further processing and marketing. For decades, large volumes of associated gas produced at remote oil fields without access to gas pipelines were simply vented or burned off (flared). This wasteful practice is still common in giant western Siberian oil fields: they also yield large volumes of associated gas, and a large share of these flows continues to be flared (Røland 2010). This environmentally damaging practice has been greatly reduced since the 1970s, but the total amount flared globally is still unacceptably high. In 2010 it was estimated at 134 Gm^3 (mostly in Russia, Nigeria, and Iran), equivalent to almost 20% of the US gas use (GGFR 2013).

The world's largest oil field, the Saudi al-Ghawar, is also an excellent example of a reservoir containing both liquid and gaseous hydrocarbons: besides producing annually about 250 Mt (10.5 EJ) of oil, it also yields about 21 Gm^3 (750 PJ) of associated natural gas (Alsharhan and Kendall 1986; Sorkhabi 2010). And in 1971, 20 years after the reservoir began producing crude oil, a large pool of nonassociated gas was discovered below the oil-bearing layers at a depth of 3–4.3 km; this deep reservoir now produces annually about 40 Gm^3 (1.4 EJ) of nonassociated gas. North America's latest addition to large formations producing both oil and gas is the Williston Basin (Bakken shale) in North Dakota, where oil extraction by horizontal drilling and fracturing has been accompanied by so much natural gas that, in the absence of adequate pipeline capacity, large volumes of it had to be flared: by the end of 2011 more than one-third of all gas produced in North Dakota was flared or not marketed (USEIA 2011b).

The third large source of natural gas is coal beds, and China and the United States, the largest coal mining nations, are also the largest producers of gas from this source. The latest addition is natural gas released by horizontal drilling and hydraulic fractioning from shales: gas-bearing shales underlie large areas on all continents, but so far only the United States has developed this resource on a large scale (Maugeri 2013). US statistics show the relative importance of these four principal gas sources: in 2011, 43% of gross withdrawals came from natural gas wells, 21% from oil wells, 6% from coal bed wells, and 30% from shale gas wells, whose output was lower than that of coal bed wells as recently as 2007 (USEIA 2014c).

Power Densities of Gas Production and Delivery

As with oil extraction, both well densities and the areas of well sites required for natural gas extraction vary, but, not surprisingly, they closely resemble each other. In conventional US natural gas fields, the typical density is one well per 256 ha, that is, 0.4 wells/km^2, but the rate is significantly higher in shale or tight gas formations: in the Barnett shale it was 1.1–1.5 wells/km^2 in the beginning, and later permits allowing in-fill wells have resulted in up to 6 wells/km^2 (NYSDEC 2011). Future spacing will be less dense as multiple horizontal wells are drilled from a single well pad. For example, in the Marcellus shale, which stretches from West Virginia to New York, a standard vertical well may be exposed to no more than 15 m of the reservoir while a lateral bore of a horizontal well can reach 600–2,000 m within the targeted formation (Arthur and Cornue 2010). This means that a producer can develop one square mile of subterranean resources with 16 vertical wells (with 40-acre spacing), or as few as four horizontal wells drilled from a single pad.

In Pennsylvania's Marcellus shale, a typical multiwell pad for drilling and fracturing is 1.6.–2.0 ha, which, after a partial restoration, may be as little as 0.4 ha; in New York State a new multiwell pad set up for horizontal drilling and fracturing is 1.4 ha, and that is reduced to 0.6 ha after the required partial reclamation, slightly smaller than a soccer field at 0.7 ha (NYSDEC 2011). The production of natural gas from shales shows the same rapid hyperbolic decline as does the extraction of crude oil (Sandrea 2012). Wells in the Marcellus shale, the most extensive gas-bearing formation in the United States, produce at an initial rate of 120,000 m^3/day, followed by an early decline of 75%, while in the Barnett shale in Texas the initial flow is less than 60,000 m^3/day, followed by a 70% decline.

The first-year flow in Pennsylvania may be in excess of 10 Mm^3 (average of almost 6 Mm^3); the next year the mean is down to less than 2 Mm^3, and that flow is halved in the third year (Harper and Kostelnik 2012; King 2013). The highest first-year flows (with a 0.5-ha pad) imply an extraction power density in excess of 2,000 W/m^2, while the third-year flow is down to just above 200 W/m^2. In western states the initial sizes of well pad sites average nearly 1.6 ha, but over the life of production that is reduced to less than 0.6 ha (Buto, Kenney, and Gerner 2010). These states, as well as the gas-producing areas of Oklahoma, Texas, and California, have hundreds of thousands of old stripper wells whose production is only 100,000–200,000 m^3/year, and even with a small well site area of 0.5 ha, their extraction power densities would be less than 50 W/m^2 (USEIA 2012).

Typical new productive wells have extraction power densities two orders of magnitude higher. In Alberta, wells claim 1.5–15 ha (average of 3 ha), and with an average annual output of about 50 Mm^3 (Jordaan, Keith, and Stelfox 2009), this means an extraction power density of about 1,850 W/m^2. Dividing Alberta's 2012 province-wide statistics on natural gas output (100 Gm^3) by the 1,622 connected wells (Alberta Energy 2013) yields an average productivity of nearly 62 Mm^3/well, and a typical power density for well sites of 3 ha would be about 2,300 W/m^2. And power densities can be another order of magnitude higher in supergiant fields.

The Groningen field is one of the best examples of minimal and unobtrusive gas recovery. The extraction is concentrated in 29 production clusters remotely controlled from the central control room in Sappemeer; each cluster has 8–12 wells arranged in a strip and adjoined by associated treatment plants (comprising several identical units), an electricity supply, and a control building (Royal Dutch Shell 2009). Clusters are very similar in size, with almost 7 ha taken by wells and 4 ha by other facilities, and the total area claimed by 20 clusters is about 310 ha. With the annual

2010–2020 production capped at 43.6 Gm^3, this puts the field's power density of extraction at about 16,000 W/m^2.

Gathering lines (low pressure, small diameter) take gas from individual wellheads to processing facilities to strip CO_2, H_2O, and also H_2S. The processing of natural gas has minimal land requirements, with power densities ranging mostly between 50,000 and 70,000 W/m^2. Purified gas is sent through high-pressure, large-diameter (0.5–1.05 m) transmission (trunk or interstate) lines. US statistics show the difference in the aggregate length of three categories of gas pipelines (PHMSA 2013). In 2012 the country had about 16,800 km of onshore gathering lines and 477,500 km of transmission lines. Compressor stations take up about 2 ha at intervals of 65–150 km, and they have slightly lower throughput power densities, on the order of 20,000 W/m^2. Distribution lines (very low pressure, diameters of just 1.25–5 cm, also as plastic tubing) take gas to individual users, and their length of more than 3.2 Gm far surpasses the combined extent of all other pipelines.

An average throughput power density for US natural gas pipelines can be calculated from detailed information about individual US systems (USEIA 2007). The Northern Natural Gas Company, which serves states from Texas to Illinois, is the country's longest interstate natural gas pipeline system, with about 25,400 km of trunk lines and an annual capacity of 78 Gm^3; with an average ROW of 15 m, this implies a throughput power density of about 230 W/m^2. Texas Eastern Transmission (14,700 km, carrying 66 Gm^3 from the Gulf of Mexico to the Northeast) rates about 330 W/m^2. Columbia Gas Transmission Company, which serves the Northeast (16,600 km), has the highest annual capacity (86 Gm^3) and a throughput power density of about 380 W/m^2. Algonquin Gas Transmission in New England (1,800 km and 24 Gm^3) rates about 1,000 W/m^2.

Liquefied Natural Gas

Increasing volumes of the fuel are traded in the form of liquefied natural gas (LNG). Gasification entails cooling the gas to $-162°C$ and reducing its volume to about 1/600th of the gaseous state; this is done in several independent units (trains) that are typically about 300 m long (Linde 2010). LNG is stored in superinsulated containers before it is transferred (via articulated pipes) to isothermal tanks on LNG tankers, the largest of which, the Q-Max ships of Qatar, carry 266,000 m^3 of gas (Qatargas 2013). With

22.2 GJ/m^3 of LNG, that equals almost 6 PJ of floating energy storage. Regasification takes place in seawater vaporizers. Most of the recently commissioned terminals have liquefaction capacities of 4–5 Mt/year, and most of the new receiving terminals rate between 3 and 5 Mt/year (IGU 2012).

By the end of 2013 more than 40 liquefaction terminals were in operation or under construction in 19 countries, while nearly 30 countries had more than 90 regasification terminals, with the largest number (24) in Japan (Global LNG 2013). Both the gasification and regasification terminals have relatively small land claims (typically 30–120 ha), and, as with oil tankers, actual LNG loading and offloading are often done offshore. A liquefaction capacity of 3 Mt/year and an area of 80 ha would translate to a throughput power density of 6,400 W/m^2. The Norwegian Snøhvit LNG plant, Europe's first world-scale LNG project, located on Melkøya island near Hammerfest, has an annual capacity of 4.3 Mt LNG, and its compact modular design occupies only about 70 ha (Nilsen 2012); that (with 53.6 GJ/t LNG) implies a processing power density of about 10,000 W/m^2. The first three liquefaction trains at Ras Laffan, the world's largest LNG exporting facility, have an annual capacity of 10 Mt and claim 3.7 km^2 (Qatargas 2013), resulting in throughput power density of about 4,600 W/m^2.

Regasification plants have power densities as high, or even considerably higher. America's largest receiving LNG terminal, Cheniere Energy's terminal on the Sabine Pass River in Louisiana, has a capacity of 19.5 Mt/year and a maximum power density of about 8,300 W/m^2. Japan is the world's largest importer of LNG, with more than 30 regasification terminals. Higashi Niigata on the coast of the Sea of Japan (annual capacity of 8.45 Mt, area of 30 ha) has a power density of nearly 48,000 W/m^2. The Himeji and Himeji Joint LNG terminals have a combined capacity of 12.67 Mt/year on 60 ha of landfill on the northern shore of Seto Inland Sea; this yields a throughput power density of nearly 36,000 W/m^2. Tokyo Bay's Futtsu terminal, the world's largest regasification facility, supplies fuel for the world's second largest gas-fired electricity-generating plant, the 4.534-GW Futtsu station (TEPCO 2012). With an annual capacity of 19.95 Mt and an area of just 5 ha (ten storage tanks holding up to 1.1 Mm^3), the terminal's throughput power density is about 60,000 W/m^2.

5 Thermal Electricity Generation

Thermal electricity generation proceeds either by burning fossil fuels or by fissioning uranium, and accounting for its space requirements is a rather complex endeavor. Direct claims for plant sites are fairly small, but indirect claims for energy inputs (including their transport), the management of wastes, and electricity transmission reduce the overall power densities.

Most of the world's electricity—almost exactly 80% in 2010—is produced by large thermal power plants burning fossil fuels or fissioning uranium. The development of thermal electricity generation began with the first Edisonian coal-fired stations of the early 1880s, which relied on large steam engines and dynamos, and the standard setup that prevails to this day—the combustion of coal in large boilers and the generation of electricity using steam turbogenerators invented by Charles A. Parsons (whose first patent was granted in 1884)—became the norm by the beginning of the twentieth century (Smil 2005). Subsequent progress was slower, and large gains in capacities and performances resumed only after World War II (Smil 2006). After 1950 an increasing share of thermal electricity was produced by burning hydrocarbons, beginning with oil and progressing to natural gas, with most of the latter fuel eventually destined for gas turbines rather than for large boilers.

In the late 1950s came the commercialization of nuclear electricity generation based on highly exothermic fissioning of uranium to generate steam in a variety of reactors. Fuel is arrayed in rods, and water can serve both as a moderator (to slow down neutrons) and as a working medium, but some designs use gas for cooling. The pressurized water reactor (PWR) eventually emerged as the dominant design: in 2013 the world had 435 operating reactors, of which 274 were PWRs (IAEA 2013). Other relatively common choices have included British gas-cooled reactors and Canadian CANDU (CANadian Deuterium Uranium) reactors using natural rather than enriched uranium.

The rapid expansion of nuclear capacities came during the 1960s and 1970s, followed by the virtual cessation of building in the United States and Canada and limited additions in Europe (mostly in France, the country with the highest reliance on nuclear generation). Since the year 2000 most new additions have been in Asia (China, Japan, South Korea, India), but the Fukushima plant disaster of March 2011 has affected those prospects. All thermal plants share key components: large halls with turbogenerators, cooling towers (in locations where there is no possibility for once-through cooling), and large transformer yards. But fossil-fueled stations require fuel storage, and all modern coal-fired plants have appropriate facilities to control air pollutants. Complete accounts of the land needed for thermal electricity generation must go beyond power plants and quantify the impacts of the requisite fuel production and its delivery. That is why my account of the power densities of thermal electricity generation is divided into the three categories that now dominate the industry: coal-fired plants, gas-fired installations, and nuclear power stations.

Coal-Fired Power Plants

Since the early twentieth century, increasing shares of global electricity have come from hydroelectric power, the burning of hydrocarbons, and, since 1956, also from nuclear fission, but coal has remained the single largest energy source for thermal electricity generation. The fuel has been particularly prominent in the world's two largest economies: in the United States it still accounted for 56% in 1990 and 51% in the year 2000, and only the recent rise of gas-fired electricity generation has brought its share to 43% in 2012 (USEIA 2014d). In China, coal-fired power plants generated about 80% of electricity in 1975, and the share was still 81% of the much-expanded output in 2012 (World Coal Association 2013).

The principal components and configurations of coal-fired power plants have not changed for decades, but the stations have larger boilers and turbogenerators (the two forming a generation unit), and large multi-unit stations have become the norm. The largest turbogenerators eventually reached ratings in excess of 1 GW, and the largest plants, made up of a series of units (boiler-turbogenerator assemblies), have capacities up to 5 GW. In light of the wide range of their designs and performances and the different origins and qualities of their fuel—the extremes would be a mine-mouth station located next to a large surface coal mine and burning poor-quality lignite, and a coastal station receiving high-quality coal shipped in bulk carriers—it is not surprising that land claims of coal-fired stations vary widely.

Principal Structures

Dominant structures that are shared by all coal-fired power plants—tall boiler buildings, machine (turbogenerator) halls, electrostatic precipitators, tall stacks, maintenance and office buildings, a switchyard, and coal storage areas—occupy a relatively small amount of space. Coal combustion in large boilers and the expansion of highly compressed steam in turbines are two energy conversions that proceed at very high power densities. Modern plants burn mixtures of air and pulverized coal (with most particles as fine as baking flour, or less than 75 µm in diameter) in a swirling vortex, with flames reaching temperatures as high as 1800°C.

Modern boilers (steam generators) are tall prismatic structures with walls covered in stainless steel tubing that circulates water to be converted to pressurized steam (Teir 2002). The boiler supplying steam to a large 1.3-GW turbogenerator is about 52 m tall, and its footprint is nearly 520 m^2 (33.8 m × 15.5 m), which means that the combustion power density will be between 6.5 and 7 MW/m^2; smaller boilers in less efficient plants have power densities of 2–4 MW/m^2. Large modern steam turbines are also very compact: for example, the 250-MW$_e$ Siemens design is 19 m long and 8 m wide, the Škoda 660-MW unit is about 32 by 13 m, and the Siemens 1,350-MW turbine is about 70 m long and 30 m wide (Fiala 2010; Siemens 2013a). The rectangular footprint of these turbines implies approximate operational power densities of 0.6–1.5 MW$_e$/m^2. The core structure of a coal-fired power plant—boilers and turbogenerator halls—thus claims only a small fraction of a station's total area: for a 1.5-GW (with three 500-MW turbines) station it can be just 1.2–1.6 ha, while a larger (2–3 GW) station powerhouse can claim 2–3 ha, resulting in specific power densities of 10^5 W$_e$/m^2.

The undesirable consequences of high-efficiency coal combustion include the voluminous generation of fly ash and (where high-sulfur coal is burned) of SO_2. That is why all modern coal-fired stations must invest into expensive techniques to limit the emissions of these air pollutants. Structures housing these air pollution control facilities are sited immediately adjacent to boilers and tall chimneys. A typical linear configuration would proceed from a coal bunker feeding a ball mill to pulverize coal, a boiler, an air heater, an electrostatic precipitator, a flue-gas desulfurization (FGD) unit, and a tall stack. In plants with capacities of 1.5–3 GW, electrostatic precipitators used to capture fly ash will occupy 1–3 ha, and FGD units will also occupy 1–3 ha. Adjacent walkways and separation spaces will raise these totals by 50%.

Captured fly ash can be used to make cement, an appealing choice because it lowers the energy cost of the material and avoids landfilling. In China, two-thirds of all captured fly ash (a total of nearly 400 Mt/year) has recently been used by the cement industry (Lei 2011). In the United States, Virginia's Ceratech blends 95% fly ash and 5% liquid ingredients to make a stronger concrete (Amato 2013). FGD removes SO_2 by reacting with lime or limestone to produce $CaSO_4$, which can be used in wallboard manufacture. Despite these efforts, large volumes of fly ash and FGD sludge are still deposited within many plant sites, claiming considerable amounts of land during the 35–40 years of typical power plant operation: for 2–3 GW plants these claims can add commonly to between 120 and 160 ha, that is, a disposal power density of 1,300–1,900 W_e/m^2. For coal with a high ash content the areas will be obviously larger, 200 ha for an Indian 1-GW_e plant burning domestic raw coal with 40% ash, and 480 ha for a 4-GW_e station burning domestic sorted coal with 34% ash (CEA 2007).

All thermal stations must condense steam discharged by the turbines and cool the condensate. This can be done by once-through cooling, which withdraws water from streams, lakes, or from the ocean with little additional space requirement; spray ponds require about 400 m^2/MW_e (2,500 W_e/m^2), and ordinary cooling ponds need up to 5,000 m^2/MW_e (200 W/m^2). Cooling towers allow water's continuing reuse. Thus, a complete water system for a 1-GW_e plant (including water treatment and cooling) could occupy up to 20 ha for natural draft towers and only 10 ha for induced draft cooling (CEA 2007). Natural draft towers have inevitable evaporation losses but require no power to operate; their power densities (per square meter of tower base) are between 20,000 and 40,000 W_e/m^2. Mechanical draft towers (counterflow or cross-flow designs) use fans to move the air, and in dry cooling towers there is no evaporation as water circulates in closed pipe circuits; the power densities of their heat rejection are high, on the order of 10^5 W_e/m^2.

Mine-mouth stations can be supplied directly from adjacent surface or underground extraction, but stations receiving deliveries by trains, barges, or ships need on-site storage for 90–120 days of normal operation. The energy storage density of coal yard will depend on the coal quality and the thickness of a coal pile (commonly 10–12 m). The area needed for storing 90 days' worth of coal for a 3-GW station will be 15–20 ha. The total area of a 3-GW station with adequate on-site coal storage and requisite ash and FGD sludge disposal will thus be between 350 and 500 ha, that is, a power density of up to 850 W_e/m^2 for installed capacity and (with a 75% load factor) up to 650 W_e/m^2 for actual generation.

Variable space requirements surpass the fixed land claims, which remain constant or change only in minor ways if there is no major reconstruction or expansion. Coal mining claims, the single largest space requirements, can be accurately quantified only for plants supplied by adjacent mines or for the stations that receive their fuel from a single distant place of origin. In reality, many plants receive fuel from different source, while others change their suppliers during the decades of their operation (in Europe the switch has commonly been from original domestic coal to cheaper fuel imported from Australia, Indonesia, South Africa, or the United States). This makes it difficult, if not impossible, to impute any definite power densities to individual plants. Generalizations are also impossible for fly ash storage (ash content can range from less than 10% to more than 30%) and for FGD wastes. Finally, there are major differences in land claims imposed to connect the stations to existing high-voltage grids. Some plants will require only short new links, others will necessitate upgrading of older lines or the construction of new connectors.

Two Realistic Examples

To indicate the actual range, I will present two realistic examples with substantially different power densities. The first case is a high-density setup: a plant with an installed capacity of 1 GW_e and operating with a high conversion efficiency of 40% will require 2.5 GW_t of coal. Its mine-mouth location means that coal can be supplied either by high-capacity conveyors or by short-haul trucking directly from the mine, obviating a large storage yard.

The plant burns good-quality bituminous coal with an energy density of 24 GJ/t, an ash content of 4%, and a sulfur content below 0.5%. It has access to a nearby source of cooling water and hence can do without any massive cooling towers. Finally, it operates with a high capacity factor of 80%. That station would generate annually about 7 TWh (about 25 PJ), and with 40% efficiency it would require about 63 PJ of coal.

I also assume that the plant's bituminous coal (energy density of 25 GJ/t, specific density of 1.4 t/m³) comes from a large surface mine whose main seam is 6 m thick and whose recovery rate is 95%; this means that under every square meter of the mine's surface are 8 t of recoverable coal containing 200 GJ of energy. To operate the plant would require annually coal extraction from an area of 31.5 ha (315,000 m²). Adding 10% for access roads, buildings, and parking lots would raise the total annual mine claim to almost 350,000 m² and would mean that the mining would proceed with a power density of about 2,300 W/m².

In this optimal case I assume that, except for the initial cut to expose the seam, the overburden would not claim additional land but would be redeposited into the mined section as the extraction front moved forward as an advancing indentation in the landscape; eventual recultivation would leave

Box 5.1
Coal for a 1-GW plant

> 1 GW × 0.8 (capacity factor) = 800 MW
>
> 800 MW × 8,766 hours = 7.0 TWh
>
> 7.0 TWh × 3,600 = 25.2 PJ
>
> 25.2 PJ/0.4 = 63 PJ

Box 5.2
Power density of coal mining

> 6 m³ × 0.95 × 1.4 t/m³ = ~8 t
>
> 8 t × 25 GJ/t = 200 GJ
>
> 63 PJ/200 GJ = 315,000 m² × 1.1 = 346,500 m² = ~350,000 m²
>
> 800 MW/350,000 m² = 2,285 W/m² = ~2,300 W/m²

behind a flat landscape approximately 6 m lower than its surroundings. In the absence of large coal storage and with once-through cooling, the largest areas occupied by the plant itself will be its generating halls, maintenance and office buildings, parking lots, switchyard, and the pond used to deposit the captured fly ash.

A generous allowance for plant structures would be on the order of 10,000 m^2, the switchyard would take no more than 50,000 m^2, and because all captured fly ash would be used in nearby cement production, there would be no need for any on-site storage. Even after adding another 40,000 m^2 for roads, walkways, parking lots, and a green buffer zone, the site's total would be 150,000 m^2, and it would prorate to about 5,300 W/m^2. After adding the coal mining area, the annual electricity generation at a rate of 800 MW would thus require about 500,000 m^2, and the power density of the entire extraction-generation sequence would be about 1,600 W/m^2 (800 MW/500,000 m^2 = 1,600 W/m^2).

The second case aggregates the factors that maximize the land claim, which results in a much lower overall power density. Again, it is a 1-GW$_e$ plant, but an older one operating with only 35% efficiency and with a lower load factor (70%). It is located far from a coal mine and is supplied by a unit train peddling between the mine and the plant. It burns low-quality subbituminous coal (18 GJ/t) extracted mainly from a 6-m-thick seam that contains 10% ash and about 2% sulfur, and it requires large cooling towers. The plant burns an almost identical amount of coal energy, but the mining

Box 5.3
Power density of a 1-GW coal-fired plant

> 1 GW × 0.7 (capacity factor) = 700 MW
>
> 700 MW × 8,766 hours = 6.14 TWh
>
> 6.14 TWh × 3,600 = 22.1 PJ
>
> 22.1 PJ/0.35 = 63.1 PJ
>
> 6 m^3 × 0.95 × 1.4 t/m^3 = ~8 t
>
> 8 t × 18 GJ/t = 144 GJ
>
> 63.1 PJ/144 GJ = 438,190 m^2 × 1.1 = 482,009 = ~480,000 m^2
>
> 700 MW/480,000 m^2 = 1,458 W/m^2 = ~1,500 W/m^2

of that fuel claims a much larger area and proceeds with a power density of only about 1,600 W_e/m^2. Adding, once again, 10% to account for land claims accompanying coal extraction lowers that rate to about 1,500 W/m^2.

The second plant would also need off-loading facilities and coal storage large enough to supply the plant for up to 60 days, a larger area for fly ash disposal, and a pond for storing slurry from FGD, and land on which to site the cooling towers. Coal for 60 days of generation would amount to 575,000 t (about 410,000 m^3) and, when stored in a yard 10 meters deep, would occupy 41,000 m^2 and up to 50,000 m^2, including approaches and off-loading ramps. Captured fly ash (with a density of 1.8 t/m^3 after compaction) deposited in a 5-m-thick layer in a settling lagoon would annually claim about 38,500 m^2 (nearly 4 ha), and FGD working with 85% efficiency would remove annually nearly 60,000 t of sulfur; when captured as $CaSO_4$ to be deposited in a 5-m-thick layer in a pond near the plant, its storage would add annually nearly 22,000 m^2.

With, again, some 150,000 m^2 for all plant structures and a switchyard, the land required for the plant's operation would add up to about 260,000 m^2, implying a power density of about 2,700 W/m^2. The plant (260,000 m^2) and coal extraction (480,000 m^2) would thus require a grand total of 740,000 m^2, and the overall power density of producing 700 MW of electricity would be about 950 W/m^2, a claim more than one-third more

Box 5.4
Power densities of fly ash and sulfate disposal

63.1 PJ/18 GJ/t = 3.5 Mt coal

3.5 Mt × 0.1% = 350,200 t ash × 0.99 (99% capture) = 346,500 t

346,500 t/1.8 t/m^3 = 192,500 m^3

192,500 m^3/5 = 38,500 m^2

3.5 Mt × 0.02% S = 70,000 t S

70,000 t S × 0.85 = 59,500 t S

59,500 t S × 4.25 = 252,900 t $CaSO_4$

252,900 t/2.32 t/m^3 = 109,000 m^3

109,000 m^3/5 = 21,800 m^2

extensive than in the first case. These two realistic constructs set a low 10^3 W_e/m^2 as the right order of magnitude for power densities of large coal plants: roughly 2,500–5,000 W_e/m^2 for the plant itself (including all structures and storages) and roughly 1,000–1,500 W_e/m^2 for a compact plant and coal mining of good-quality fuel.

Land claims are greater for plants burning low-quality coal and requiring extensive coal- and fly ash–handling arrangements. A report on land requirements of India's new coal-fired plants (burning domestic fuel containing 15 GJ/t and up to 40% ash) illustrates these needs for a typical 1-GW_e (two 500 MW_e units) station (CEA 2007). The plant buildings would occupy only 12 ha, but the total area needed would be 240 ha (with coal handling and a surrounding green belt accounting for about 60% of that total); fly ash storage (18 m high, sufficient for 25 years) would alone claim 200 ha, raising the overall claim (excluding coal mining) to more than 500 ha and resulting in a power density of about 150 W_e/m^2 for generation at 75% of installed capacity. A 4-GW_e Indian station would claim about 1,000 ha and would have a generation power density of about 300 W_e/m^2,

Actual Power Densities

Coal-fired power plants often occupy substantially larger areas than those outlined in the preceding calculations or in typical recommendations. This comes about for two main reasons: many sites keep fenced-in land in reserve for possible future expansion, including additional land that might be needed for fly ash and FGD sludge disposal, and most sites contain green areas (groves, grasses, wetlands) to buffer the plant operation and make its presence environmentally more acceptable. As a result, areas within a station's perimeter fence are, particularly in the land-rich United States, often two or three times as large as the land presently claimed by the plant's structures and storages.

The world's largest coal-fired power plant, with a capacity of 5.824 GW_e, is on about 300 ha of reclaimed land on the eastern shore of the Taiwan Strait near Taiwan's city of Taichung (GIBSIN Engineers 2006). The station burns coal imported from several countries (mainly Australia), and with an annual generation of 42 TWh its power density (excluding that consumed in coal extraction) is about 1,600 W_e/m^2. Two of America's three largest coal-fired electricity-generating plants belong to the Southern Company, headquartered in Atlanta: the Robert W. Scherer (four units, 3.52 GW), located on the northern shore of Lake Juliette southeast of Atlanta, and Bowen (four units, 3.499 GW), located in Bartow County, Georgia (fig. 5.1). Scherer is a sprawling installation with a large oval-shaped coal storage yard (36 ha), an ash-settling pond of 120 ha, and an ash disposal pond of 300 ha (designed to last for the plant's life span of some 50 years), and the plant's total operating area covers about 14 km^2 (Georgia Power 2013).

With an average load factor of 75%, the plant generates 23.1 TWh/year, and the power density of its operation is about 190 W/m^2. But the entire property, including part of Lake Juliette, is 34 km^2. The Bowen plant claims about 3.7 km^2 (with main buildings, cooling towers, the coal yard, and the transforming area accounting for only about 10% of the total area), and its annual generation of 22.6 TWh translates into an overall power density of nearly 700 W$_e$/m^2 The third largest US coal-fired plant is Duke Energy's Gibson, in Indiana (five units, 3.34 GW). The plant owns about 16 km^2 of land, of which 12 km^2 is a man-made lake used as the plant's cooling pond; the plant itself and its associated coal storage and pollution-control infra-structures take more than 2.5 km^2, resulting in a power density of close to 1,000 W$_e$/m^2. The Tennessee Valley Authority's Bull Run station (with a single 900-MW$_e$ unit generating 6 GWh/year) uses once-through cooling and occupies only 65 ha (for a power density of about 1,050 W/m^2).

Bełchatów, Poland's and Europe's largest coal-fired plant (5.298 GW$_e$, 28 TWh annual generation), burns lignite from an adjacent surface mine, and its rather large area results in a generation power density of about 500 W$_e$/m^2; the entire mining-generation system rates less than 250 W$_e$/m^2. Drax, England's largest thermal station, has a capacity of 3.96 GW, and in 2012 it generated 27.1 TWh; that implies a load factor of 78%, or 3.09 GW (Drax Group 2013). The plant owns 750 ha, but its structures and storage facilities cover less than 200 ha (with the rest of the site under trees and meadows), resulting in an actual generation power density (exclusive of coal mining) of about 1,500 W$_e$/m^2.

Jänschwalde, Germany's largest lignite-fueled station, near Peitz in Bran-denburg (very close to the Polish border), occupies about 220 ha, has an installed capacity of 3 GW$_e$ (six 500 MW$_e$ units) and generates 22 TWh/year (Vatenfall 2013). The site's power density is thus about 1,100 W$_e$/m^2. And the hard coal–fired Westphalen plant near Hamm became the world's most efficient producer of its kind with the addition of two 800-MW units that replaced two smaller units from the 1960s (RWE 2014); the new units

Figure 5.1
Scherer (top) and Bowen power plants, Georgia, United States. Both photos available at Wikimedia.

claimed only 11 ha, the entire plant occupies about 70 ha, and it has a power density of more than 2,500 W_i/m^2 (fig. 5.2).

The two best conclusions are that when the calculations include only the area actually occupied by plant's structures, then most coal-fired power plants generate electricity with power densities in excess of 1,000 W_e/m^2, and that the inclusion of coal extraction lowers those rate by widely differing margins. Power densities of the entire coal-to-electricity sequence can stay well above 1,000 W_e/m^2 but can be reduced by an order of magnitude. For comparison, Mielke's (1977) average US rate (including that consumed in coal extraction) for uncontrolled generation was nearly 1,000 W_e/m^2, 2,700 W_e/m^2 for emission-controlled western coal-based plants and less than 700 W/m^2 for plants burning eastern coal with air pollution controls.

Hydrocarbons in Electricity Generation

The consumption of hydrocarbons in electricity generation has seen two shifts: away from liquid fuels (mostly fuel oil in large power plants and diesel fuel in smaller generators) to natural gas, and from gas combustion in large boilers to the widespread deployment of gas turbines (Smil 2010b). As world oil prices rose during the two consecutive OPEC-driven energy "crises" (1973–1974 and 1979–1981), liquid fuels became too expensive to be used for electricity generation, and most of the countries either shut down them down or converted them to burning either coal or natural gas. US statistics show the rapidity of this shift: in 1980, 11% of electricity came from petroleum fuels; a decade later the share was down to 4%, and in 2012 it was only about 0.5% (USEIA 1993, 2014d).

Natural gas has made up the difference. In 1990 US utilities produced only 9% of electricity from natural gas, but a decade later the share had nearly doubled, to 17%, and in 2012 it reached 30%. Not surprisingly, the world's largest operating station burning fuel oil is in Saudi Arabia, but Japan's Kashima (4.4 GW_e burning heavy oil and crude oil) surpasses Russia's Surgut-1 plant in west-central Siberia (3.268 GW_e). The Saudi Shoaiba plant (installed capacity of 5.6 GW_e), on the Red Sea about 100 km south of Jeddah, is also the kingdom's largest plant; it now operates 14 generating units and combines electricity generation with desalination (Alstom 2013). The plant itself occupies about 250 ha, implying a power density of about 2,200 W_i/m^2.

Figure 5.2
Westphalen power station, Germany. © Hans Blossey/imagebroker/Corbis.

Kashima, located on about 60 ha of reclaimed land on the Pacific coast northeast of Tokyo, rates more than 7,000 W_i/m^2, and Surgut-1 rates about 5,400 W_i/m^2. One of the most unusual oil-fired stations was the Chavalon plant at Vouvry, Switzerland (Chavalon 2013): it had a capacity of 300 MW_i, it operated between 1965 and 1999, and its compact design—occupying just 3.6 ha on a small plateau on a steep slope above the Rhône valley just south of the eastern end of Lac Léman—resulted in a power density of more than 8,000 W_i/m^2 (its replacement, a gas-fired combined-cycle station of 400 MW_e, was approved for the site in 2010).

Natural Gas in Electricity Generation

The typical large central power plant fueled by natural gas and supplied by a pipeline claims much less land than other fossil-fueled operations because it has only minimal emergency fuel storage and no need for fly ash or SO_2 capture. When new construction of such stations was fairly common in the United States, Mielke's (1977) data put their average power density at about 1,800 W_e/m^2, and six years later the USDOE (1983) estimated that an 800-MW_e plant with a 55% load factor would occupy about 36 ha, implying an operating power density of about 1,200 W_e/m^2. Currently the largest group of high-capacity (more than 2 GW_i) natural gas–powered stations is in Japan, a country that imports 100% of its natural gas, a choice dictated by the quest for high air quality in Japan's densely populated urban regions.

Japan has nearly 20 liquefied natural gas (LNG)-based high-capacity power plants, usually located on reclaimed coastal land and close to an LNG terminal, with boilers and generator halls clustered around two or more tall chimneys and with adjacent gas storage tanks and security land buffers. The largest plant is TEPCO's Futtsu, in Chiba prefecture, rated at 5.04 GW_e, with an adjacent LNG terminal and ten storage tanks on the eastern shore of Tokyo Bay (fig. 5.3). Japan's second largest natural gas–fired station is Kawagoe (4.8 GW_e, about 30 TWh/year, with six large gas storage tanks), in Mie prefecture (Chubu Electric Power 2013). These are highly compact facilities: Kawagoe occupies only about 75 ha, Futtsu about 125 ha, and hence their operating power densities are 2,800–4,500 W_e/m^2.

Other countries with high-capacity natural gas–fired stations include Australia, Malaysia, and Russia (all burning abundant domestic gas) and China, Taiwan, and South Korea (relying on imported LNG). In 2013 Russia's Surgut-2, in Khanty-Mansyisk region of west-central Siberia,

Figure 5.3
Futtsu power station and LNG terminal, Japan. Tokyo Electric Power Company
(TEPCO).

was the world's largest natural gas–fired station (installed capacity of
5.597 GW_e, annual generation of roughly 40 TWh); with an area of about
80 ha its power density was approximately 6,000 W_e/m^2. Ravenswood in
Long Island City (Queens, New York) burns natural gas, fuel oil, and kerosene to power units with a total capacity of 2.48 GW. The plant is situated
on a small (just 12 ha) rectangular lot just south of the Roosevelt Island
Bridge on the East River, and its power density is about 20,000 W_i/m^2.

Gas combustion in large boilers of central power plants has been eclipsed
by burning in gas turbines, efficient and flexible energy converters that
now come in capacities ranging from 1 MW to 375 MW, the record rating
by 2013, for the Siemens SGT5–8000H model installed in Irsching near
Ingolstadt (Siemens 2013c). The Swiss Brown Boveri Corporation pioneered
the use of gas turbines just before World War II, but the widespread adoption of these machines came only during the 1960s, especially after the
great November blackout in the northeastern United States in November
1965 showed the need for swiftly deployable generators (Smil 2010b). The
expansion was also helped by the adoption of a combined gas cycle: hot

exhaust gases leaving the turbine are used to produce steam for an attached steam turbine, and the overall conversion efficiency is as high as 60%, a rate unrivaled by other modes of thermal electricity generation.

Aeroderivative turbines (jet engines adapted for stationary uses), made by GE, Rolls Royce, and Pratt & Whitney (P&W), are also very efficient (about 40%), and some are available as fully assembled units on trailers. A modified P&W FT8 jet engine with a capacity of 25 MWe fits on two trailers and can generate eight hours after arrival (PW Powersystems 2013).The turbine itself occupies no more than 140 m^2, and even with its control trailer, access roads, fuel and electricity connections, and a safety perimeter buffer it claims only 600 m^2 (a 40-m × 15-m rectangle), implying a power density of nearly 42,000 W_i/m^2. A larger unit, the 60-MW SwiftPac, is placed on concrete foundations, claims less than 700 m^2 (85,000 W_i/m^2), and could be ready to generate in just three weeks.

Gas turbines can thus easily be accommodated on small lots in industrial areas. The Delta Energy Center in Pittsburg, California, demonstrates this flexibility (Calpine Corporation 2013a). Calpine Corporation located the plant on an undeveloped 8-ha lot at the Dow Chemical Company facility, and with an 835-MW_e combined-cycle capacity its rated power density is nearly 10,500 W_i/m^2. Gas turbines can be also placed on land that is already occupied by large power plants, a choice that eliminates contentious application and approval processes for new plant sites. The large (1.36-GW_e) English Didcot-B plant in Oxfordshire is a perfect example of this option. It was built between 1994 and 1997 within a larger preexisting site of Didcot-A, a 2-GWe coal-fired station whose construction (completed in 1968) had met with a great deal of opposition. The gas-fired plant, with more than two-thirds of the original coal plant's capacity, now takes up less than 10% of the entire site.

Nuclear Generation

I called the worldwide achievement in post-1956 nuclear electricity generation a successful failure (Smil 2003)—successful because the fission supplied about 13% of the world's electricity by 2013, or more than was contributed at that time by hydroelectricity after 130 years of developing water turbines and building large dams; a failure because in the UK, and even more so in the United States, the two countries that pioneered its introduction, nuclear generation has fallen far short of the early expectations that it would become the dominant (if not the only) way of electricity supply before the year 2000 (Smil 2010a). Will this great pause—never entirely global, because Japan kept building until the Fukushima disaster of March 2011, and construction continues and expansion plans remain in place for China, India, and Russia—be followed by a long-awaited nuclear renaissance?

Whatever the outcome, the high power density of nuclear fission has been one of its major appeals. The conversion of nuclear energy to heat proceeds at power densities (used throughout this book to mean power per unit area, not as used by nuclear engineers to mean power per unit of reactor volume) ranging from 50 to 300 MW_t/m^2 of a reactor's footprint, that is, up to an order of magnitude higher than the power densities for boilers burning fossil fuels. In addition, nuclear stations do not require any extensive fuel-receiving and fuel storage facilities, they have no need for air pollution controls and for land set aside for the storage of captured waste products, and radioactive wastes stored temporarily on-site occupy only small areas. Well-known breaches of reactor containments (Chornobyl in 1985, Fukushima in 2011) should not obscure the fact that the operation record of nuclear stations in the two countries with the largest number of commercial reactors, the United States and France, has been excellent.

In common with all other thermal plants, nuclear stations have machine halls (housing turbogenerators, steam condensers, waste heat rejection systems, and requisite pumps) and auxiliary buildings (containing water and waste treatment facilities, fuel and maintenance stores, and administrative areas), and road and rail access. Similarly, some nuclear plants use once-through cooling, but many have tall cooling towers, and all of them must have appropriately designed switchyards needed to step up the voltage before transmitting the generated electricity to a national grid. And, like other thermal plants, nuclear stations vary greatly in their overall land claims.

In 2012 almost 60% of America's total nuclear capacity (total of 118 reactors in 74 plants) was located on sites of 200–800 ha, with a modal area of 200–400 ha, of which 20–40 ha were actually disturbed during the plant's construction (USNRC 2012a). But the extreme land claims range over two orders of magnitude, with the largest areas taken up by cooling reservoirs, artificial lakes, or extended buffer areas. California's San Onofre (2.586 GW in three reactors on the Pacific coast in Oceanside, cooled by ocean water)

Figure 5.4
San Onofre nuclear power plant, California. © Corbis.

occupies just 34 ha (fig. 5.4), while the William B. McGuire station near Charlotte in North Carolina (2.36 GW) is situated on a site of 12,000 ha.

Power Densities of Nuclear Generation

San Onofre's power density prorates to about 7,600 W_e/m^2 in terms of installed capacity, and densities between 2,000 and 4,000 W_e/m^2 are common. At the other end of the size spectrum there are power densities of less than 100 W_e/m^2: McGuire rates at just 20 W_e/m^2, Wolf Creek in Kansas (1.17 GW, 3,937 ha, most of it a large cooling lake) at 30 W_e/m^2. Insofar as most of the world's reactors belong to just two dominant types (the most common being the pressurized water reactor, followed by the boiling water reactor) and their installed capacities range mostly between 400 and 1,200 MW_e, it is hardly surprising that the power densities of nuclear electricity stations in other countries are very similar.

The world's largest station, Japan's Kashiwazaki-Kariwa (like all of the country's plants, it was closed in 2011 after the Fukushima disaster, but some of its reactors are to be restarted) had an originally installed capacity of 7.965 GW in seven reactors located within a 420-ha area along the coast of the Sea of Japan in Niigata Prefecture (Sakai, Suehiro, and Tani 2009). That would be about 1,900 W_e/m^2, but a large part of that area is the surrounding wooded buffer zone, and there is also a broad green strip between two groups of reactors. Plant structures and infrastructures take up less than 200 ha, and the plant's highest output of 60 TWh in 1999 (before it began to experience operation problems) implied an operating power density of at least 3,500 W_e/m^2.

In contrast, the now infamous Fukushima Daiichi—hit by the March 2011 tsunami and the subsequent crippling of four of the six reactors and the immediate release of radioactivity into the atmosphere and later also into the Pacific Ocean—is located on a much larger piece of costal property; with about 350 ha and an installed capacity of 4.698 GW, its predisaster power density was 1,300 W_i/m^2. The plant's twin, Fukushima Daini, 12 km south of the crippled station, has a capacity of 4.4 GW in four reactors on a much smaller site of 147 ha, rating about 2,900 W_i/m^2.

Gravelines, the largest of 21 French stations and the world's fifth largest nuclear plant (with 5.7 GW in six 900-MW reactors, producing about 38.5 TWh/year) occupies only about 90 ha just west of Dunkerque on the Pas de Calais coast; its power density is thus about 6,300 W_i/m^2. Cruas on the Rhône in southern France (3.6 GW in four 900-MW units) claims 148 ha, for an installed power density of some 2,400 W_e/m^2. My last example is the Swiss Beznau, in 2013 the world's oldest operating PWR plant, with two 365-MW reactors: the plant occupies only 20 ha on an island in the Aarr river, rating about 3,600 W_i/m^2.

Land occupied by the structures and infrastructures of most nuclear stations translates into installed power densities of 2,000–4,000 W_e/m^2, and the inclusion of immediately adjacent buffer zones (which do not prevent agriculture or forestry but exclude any permanent habitation) reduces that rate by 40%–60%. And, unlike other thermal stations—some of which, like New York's Ravenswood in Queens, are not only within cities but relatively close to city centers—nuclear station are preferentially and deliberately located in areas of lower population density. In the United States more than half of all nuclear plant sites have an 80-km population density of less than 80 people/km², and more than 80% have an 80-km density of less than 200 people/km² (USNRC 2012a).

As must be expected, land claims for the uranium cycle will significantly lower the densities of the entire fuel generation-disposal sequence.

Commercially extracted uranium ores contain as little as 0.15% and as much 20% of the heaviest of all metals (WNA 2010). Their mining is followed by milling (on-site or in a nearby facility) to produce U_3O_8 concentrate (yellowcake), which contains more than 80% uranium: about 200 t of the concentrate are needed to produce the fissionable fuel required to run a 1-GW_e plant for a year. U_3O_8 is refined to UO_2 (this unenriched oxide fuels Canada's CANDU reactors) and is then converted to UF_6. This gas undergoes isotopic separation (centrifugal or by gaseous diffusion) to produce enriched UO_2, and this oxide is formed into ceramic pellets and encased in fuel rods.

The typical progression (nuclear fuel chain material balance) from mining uranium ore to fabricating the enriched fuel into reactor rods results in a mass reduction of two orders of magnitude. For electricity generation in a 1-GW_e plant operating at full capacity and consuming fuel derived from ore with 0.2% uranium by weight, the sequence is as follows (IAEA 2009): about 108,000 t of uranium ore (containing 217 t of uranium) are processed in a uranium mill to yield 245 t of U_3O_8 (containing nearly 208 t of uranium), and that compound gets converted to about 306 t of natural UF_6; enrichment produces about 38 t of enriched UF_6, and the hexafluoride is then converted to almost 29 t of UO_2 (containing 25.4 t of uranium), which is fabricated into fuel rods. This reactor fuel will generate 8.765 TWh/year (averaging about 2.9 t U/TWh) and produce 28.8 t of spent fuel.

Uranium Mining and the Nuclear Cycle

All pre–World War II uranium production (Jáchymov in Bohemia, Colorado, Cornwall, the Congolese Katanga) was done underground, and a deep mine, Cameco's McArthur River Mine in northern Saskatchewan, is the world's largest uranium-producing operation (fig. 5.5). The mine opened in the year 2000, and by 2012 its cumulative extraction (by indirect methods, using freezing and raise bore mining to minimize exposure to radioactive surroundings) reached about 104,000 t and its annual output of about 7,500 t was 13% of the global total (Cameco 2013b). The mine occupies a compact main area of about 1,000 by 400 m, four smaller outlying areas, and an airstrip facility—altogether about 100 ha, which means that only about 10 m^2 have been required per tonne of output during 13 years of operation.

Figure 5.5
McArthur River uranium mine, Saskatchewan. © STRINGER/CANADA/Reuters/ Corbis.

As 1 t of natural uranium can generate 42.2 GWh of electricity, the mine's 2000–2012 output produced 4.38 PWh of electricity (roughly 38.5 GW), that is, just 0.023 ha/TWh, and at a very high extraction power density of about 38,000 W_e/m^2 of surface disturbance. The mine's ore is milled at Key Lake, about 80 km southwest of McArthur River, at a former surface mine whose cumulative 1983–2002 production reached about 95,000 t of uranium between 1983 and 2002 and whose total disturbed area (pits, leaching ponds, tailings, roads) amounts to about 15 km². By using the metal output for both sites (440,000 t in 30 years between 1983 and 2012) and their total disturbed area (some 16 km²), we can calculate a long-term (1983–2012) average of land claims that combines mining (surface and underground) and milling at the world's two premiere uranium deposits: it has averaged less than 40 m² (0.0036 ha)/t during the three decades, and its extraction plus milling power density has been about 4,400 W_e/m^2.

But this is an exceptionally high rate because northern Saskatchewan deposits have an extraordinarily high concentration of uranium: at 16.36% U_3O_8, McArthur River ore grade is two orders of magnitude above the global average. The Olympic Dam mine in southern Australia is the world's second largest uranium producer from ore (a uranium content of 0.12%) and also yields copper, silver, and gold (WNA 2013a). Its sprawling operations cover about 21 km^2 and its cumulative 1988–2012 uranium output was about 65,000 t (WNA 2012): if the entire land claim is attributed to uranium mining, the rate is about 0.03 ha/t, or (at roughly 12 GW) 600 W_e/m^2. The Ranger mine in Australia's Northern Territories (the world's third largest uranium enterprise) produced the metal by surface mining between 1981 and 2012. Its operations yielded about 100,000 t and its pits, ponds, ore stores, tailings, buildings, and roads covered about 900 ha (with about 420 ha disturbed). The overall operation claim prorated to about 0.009 ha/t, that is, a power density of roughly 1,700 W_e/m^2.

The large open pit and associated mine structures of the Rössing deposit in Namibia, the number six producer globally, cover nearly 25 km^2 and the operation produced about 103,000 t of uranium between 1976 and 2012 (WNA 2013a), requiring about 250 m^2/t, or less than 550 W_e/m^2. But the share of surface mining has been declining as in situ leaching (ISL) has emerged as the most important method of uranium production. In 2012 that process supplied 45% of all uranium, followed by about 28% from underground seams and 20% from open pits, with the small remainder being a by-product of other extraction (WNA 2013b).

In large sheetlike deposits ISL is deployed as a gridded well field where injection wells (used to introduce an acid, or alkaline, leaching solution into an aquifer) alternate with extraction wells (from which submersible pumps lift the leachate to the surface) at a spacing of 50–60 m. Additional wells are drilled above and below the aquifers within the well field and around its perimeter to monitor the containment of the leaching solutions (IAEA 2005). In narrower, channel-type deposits the well spacing is as close as 20–30 m. Recovery rates are 60%–80% of the metal present in the deposit during 1–3 years of injection and withdrawal.

Kazakhstan, now the world's leading producer of uranium, has the largest ISL operations, but there is an increasing number of smaller projects in Australia and the United States. Data from these three countries (IAEA 2005; McKay and Miezitis 2001; WNA 2012a) show annual land requirements on the order of 0.1 ha/t of uranium. Crow Butte in Nebraska has an original license covering 1,320 ha, but the land affected the by mine's structures and 11 well fields has been only 440 ha, and the project yielded about

3,800 t of uranium between 2002 and 2012 (USNRC 2013); this translates to an average claim of about 0.11 ha/t in a decade and an extraction power density of only about 380 W_e/m^2. Australia's Beverly mine contains deposits of about 16,300 t of uranium, of which 10,600 t are recoverable by ISL below an area of 800 ha (McKay and Miezitis 2001). With an annual output close to 1,000 t of uranium (recovery in 10–12 years), this implies an extraction power density of 530–640 W_e/m^2.

Underground and surface mining and ore milling leave behind voluminous tailings, but the processing of yellowcake claims a minimal amount of space. Cameco's Canadian conversion facility in Port Hope, Ontario, is licensed to convert U_3O_8 into 12,500 t of UF_6 and 2,800 t of UO_2 (used to fuel Canada's CANDU reactors) but occupies only 9.6 ha on the northern shore of Lake Ontario (Cameco 2013a; Senes Consultants 2009). This means that even if it worked at half its annual capacity, the facility's conversion power density would be on the order of 10^5 W_e/m^2. The fabrication of fuel rods to reactor specifications is a similarly highly concentrated process with negligible space requirements.

Additional space should be allocated because of the relatively high electricity requirement of the enrichment process. This depends on the degree of enrichment, an effort measured in separative work units (SWU). The gaseous diffusion process, the original separation method, is highly energy-intensive (about 2.3 MWh/SWU), while gas centrifuge plants, now the dominant way of enriching the fuel in the United States, need only about 60 kWh/SWU (FAS 2013). The previously traced nuclear fuel chain material balance would require more than 116,000 SWU and the gaseous diffusion would need 267.7 GWh to enrich enough fuel to operate a 1-GW_e reactor at full capacity for a year. If the enriched fuel were supplied solely by gaseous diffusion, then the total annual electricity consumption would be no less than 240 GWh; if it were supplied only by centrifuge plants, the requirement would be as low as 7 GWh.

If all the fuel were supplied by a combination of the two processes in one country, and if all the requisite electricity were to originate from a single source, it would be easy to calculate a weighted mean. But the United States relies on foreign enrichment services to fuel its nuclear power plants: in 2012, owners and operators of America's commercial nuclear power reactors purchased enrichment services totaling 16 million SWU (USEIA 2013d); the electricity used to enrich it came from different (and gradually changing) national mixtures of sources; and their power densities (as we have seen in preceding sections) range over several orders of magnitude, from large hydroelectric to natural gas–fired generation.

Consequently, to do as Lovins (2011b) has done, that is, to select gaseous separation as the only choice, estimate that about 10 TWh of electricity are needed to separate the isotopes during the 1-GW plant's four decades of operation, and then assume that all of that energy comes from coal-fired power plants whose annual land requirements add up to 580 ha/TWh, is a questionable procedure. Its outcome would add about 150 ha/year to land claims of a 1-GW$_e$ nuclear station. In contrast, Fthenakis and Kim (2009) put enrichment's land requirements at about 3 m^2/GWh (assuming 70% centrifugal and 30% diffusion enrichment); for a 1-GW$_e$ plant with a 90% capacity factor that would translate to roughly an additional 2.6 ha/year, less than 2% of Lovins's huge total. Even at double that rate (about 5 ha/year), this would be a small addition that would have only a marginal effect on the aggregate count.

During the 1970s, the decade of the record expansion of US nuclear capacities, Mielke (1977) put the claims of fissile fuel production (all pro-rated for one year of a 1-GW$_e$ light water reactor operation) as follows: temporarily committed land for mining, 22 ha; for milling, 0.2 ha; for UF$_6$ production, 1 ha; for uranium enrichment, 0.2 ha; and for fuel fabrication, just 800 m^2. The corresponding totals for actually disturbed area were 6.8, 0.1, 0.08, 0.08, and a mere 160 m^2. Mielke considered only direct land claims of enrichment, but otherwise his numbers convey well the relative land demands of nuclear generation: differences for mining claims among underground, surface, and ISL operations will in most cases be far more important in determining the overall requirements than will the aggregates of all postmining operations.

Finally, there is the matter of the long-term storage of spent fuel. The fuel is removed from reactors to adjacent storage ponds, where it can stay for months or years as its radioactivity decreases; this common practice creates no additional land claims beyond plant boundaries. Cooling ponds now contain about 90% of the world's 270,000 t of all used fuel, with the remainder in dry storage (WNA 2012). In the United States, about a quarter of all used fuel is in interim storage in sealed steel casks or modules at Independent Spent Fuel Storage Installations (USNRC 2012b). The relatively limited mass and volume of these wastes help to explain why there has been no urgency to set up a permanent disposal site (the NIMBY syndrome is another consideration).

While there is still no permanent national facility for the long-term storage of highly radioactive waste, many years of planning for the Yucca Mountain project in Nevada (now essentially terminated) offer relevant insights into the storage capacities and potential land claims of such depositories (Boisseau 2009). The site's intended capacity was 70,000 t of radioactive waste, and the deposit's footprint would have been 4.27 km^2, but its controlled access area would have extended over 230 km^2 (Cochran 2008). Using the latter total, the rate would have been about 3,300 m^2/t, and hence the waste from a 1-GW_e reactor (nearly 29 t of spent fuel a year) would have an overall land claim of roughly 9.5 ha/year.

Three Representative Plants

As I did for coal-fired plants I will present two very different but realistic examples of aggregate power densities of nuclear electricity generation, as well as an example of a more commonly encountered facility, all for a standard 1-GW_e station with a high (90%) capacity factor. The first plant is a compact operation (much like San Onofre) that occupies just 50 ha and whose fuel comes from the world's most productive Saskatchewan mines (just 40 m^2/t U). The fuel is enriched only by centrifugal process (6 GWh/year), with electricity supplied by coal-fired generation operating with a high power density of 1,000 W_e/m^2 (that is, 8.76 MWh/m^2). Spent fuel is eventually stored in a permanent depository whose land claim amounts to 10 ha/year. The plant's land claim is dominated by its site, and it operates with a power density of about 1,600 W_e/m^2.

The second station is a sprawling enterprise that spreads over 1,000 ha and receives uranium through a low-density ISL recovery (0.2 ha/t U). Enrichment of the uranium is split between a gaseous diffusion process (about 268 GWh needed to enrich fuel for 1 GW_e) and a centrifugal process (6 GWh), resulting in an average of 137 GWh required to produce the fuel for one year of the plant's operation. Land requirements may increase even further if that electricity originated in hydrogenation operating with a very low power density of 5 W_e/m^2 (43.8 kWh/m^2). This plant's land claim is also dominated by its site, but fuel production will add another 30%, and the plant will operate with a power density of only about 70 W_e/m^2.

Box 5.5
Power density of a nuclear electricity generation (high-density variant)

Power plant site	50 ha
Generation	1 GWe × 8,760 = 8.76 TWh
Uranium requirement	217 t U/year
Mining and milling	217 t U × 0.004 ha/t U = 0.87 ha/year
Land for enrichment	6 GWh/8.76 MWh = ~680 m2/year
Spent fuel storage	10 ha/year
Total land requirement	50 + 0.87 + 0.07 + 10 = ~61 ha/year
Power density	1 GWe/610,000 m2 = ~1,600 We/m2

Box 5.6
Power density of a nuclear electricity-generating plant (low-density variant)

Power plant site	1,000 ha
Generation	1 GWe × 8,760 = 8.76 TWh
Uranium requirement	217 t U/year
Mining and milling	217 t U × 0.2 ha/t U = ~43 ha/year
Land for enrichment	137 GWh/43.8 kWh = ~310 ha/year
Spent fuel storage	10 ha/year
Total land requirement	1,000 + 43 + 310 + 10 = ~1,360 ha
Power density	1 GWe/13.6 Mm2 = ~70 We/m2

The most representative of the three nuclear power plant examples is a station whose site occupies an area right in the middle of the US modal range (200–400 ha). Other assumptions are as follows: the station receives uranium from several sources, including overseas imports; uranium extraction (with a significant share coming from ISL) and ore milling land claims average 0.1 ha/t; and the weighted fuel enrichment cost (80% centrifugal, 20% gaseous process) is about 58 GWh, with electricity coming from a mixture of sources with an average power density of about 500 W_e/m^2 (4.38 MWh/m^2). This plant's land claim is also dominated by its site,

Box 5.7
Power density of a nuclear electricity-generating plant (medium variant)

Power plant site	300 ha
Generation	1 GWe × 8,760 = 8.76 TWh
Uranium requirement	217 t U/year
Mining and milling	217 t U × 0.1 ha/t U = ~22 ha/year
Land for enrichment	58 GWh/4.38 MWh = ~1.3 ha/year
Spent fuel storage	10 ha/year
Total land requirement	300 + 22 + 1.3 + 10 = ~333 ha/year
Power density	1 GWe/3.3 Mm2 = ~300 We/m2

but fuel production and disposal add roughly 50% to the plant's area, and the station generates electricity with an overall power density of about 300 W_e/m^2.

These three realistic examples thus span annual claims of roughly 60–1,360 ha and power densities of 70–1,600 W_e/m^2, attesting again to the fact that the power densities of thermal electricity-generating stations are highly constrained as far their core structures and indispensable infrastructures are concerned but can differ by two orders of magnitude mainly because of differences in fenced-in areas (including undisturbed land), cooling arrangements, and the origins of the fuel supply.

In closing, here are a few values offered by several studies as averages for the land requirements of nuclear electricity generation. Gagnon, Bélanger, and Uchiyama (2002) estimated direct land requirements at 0.5 km^2/TWh, that is, about 440 ha for a 1-GW_e plant, and a power density of about 230 W/m^2. Fthenakis and Kim (2009) concluded that US nuclear generation claims 119 m^2/GWh; that implies a total claim of about 104 ha for 1 GW_e and a power density of roughly 960 W_e/m^2. The largest component (42% of the total) of their account was the plant itself, but it is obvious that with 42 ha (an area smaller than that controlled by the most compact US nuclear station) they counted only the footprint of its structures, not the total fenced-in area of the facility.

Their relatively high mining estimate results from the assumed split of extraction methods (50% open pit, 50% underground, no ISL), and their indirect claim for fuel enrichment reflects the 70/30 split between centrifugal separation and diffusion. Lovins (2011b) arrived at an almost identical rate of roughly 120 m^2/GWh. McDonald and co-workers (2009) offered a fairly narrow range, 3.02 km^2/GW for the most compact and 4.78 km^2/GW for the least compact nuclear plant, implying power densities of 210–330 W$_e$/m^2. These claims, 302–478 ha for a 1-GW$_e$ station, are based on a study by Spitzley and Keoleian (2004), whose assumptions were also cited by Lovins (2011b). Jacobson's (2008) calculations for an 847-MW reactor (which included all land for uranium production and a safety zone) ended up with 150 W$_e$/m^2.

Transmission

I could have quantified the land requirements of electricity transmission in the third chapter while assessing the power densities of renewable energies, but placing them here is more apposite. After all, for more than 130 years (Edison's first power plant began operating in 1882) most of the world's electricity has been generated by burning fossil fuels, and since the late 1950s also by nuclear fission: in 2014 thermal electricity generation supplied nearly 80% of the world's total demand—and technical, infrastructural, and economic imperatives make it certain that this primacy will continue for decades to come.

Much like thermal electricity generation, electricity transmission retains its basics (now more than a century old), but its performance has been much improved as better transformers, higher voltages, taller towers, better wires, and longer unsupported spans have kept pace with the rising demand for larger transfers across longer distances in more difficult terrain.

Early choices are also reflected in different consumer voltages (100 V in Japan, 120 V in North America, 230 V in Europe), but these do not affect the land claims of high-voltage transmission. Distribution lines to consumers have voltages less than 35 kV, and in modern urban developments they are underground. High-voltage transmission of alternating current (AC) operates at 110 kV and 230 kV, and it steps up to extra-high voltage at 345 kV, 500 kV, and 765 kV. Direct current (DC) transmission, a more efficient alternative to connect load centers (cities, industries) with distant sources of power (usually large hydroelectric plants), has higher voltages to reduce transmission losses, but the capacities of high-voltage lines are limited by their heating and by voltage drop.

The width of transmission rights-of-way increases with voltage, but extra-high-voltage lines make lower specific claims. In the United States the National Electric Safety Code specifies minimal clearances of 36–46 m for 230-kV lines, 46 m for 345-kV links, and 61 m for 765-kV (all AC) transmission (IEEE 2012). This means that 1 km of high-voltage transmission claims 3.6–6.1 ha of ROW, and the power densities of annual electricity throughput in the United States in 2012 (440 GW) would have been—with 305,000 km of lines 230 kV and higher (USEIA 2014b) and assuming an average ROW of 5 ha/km—just short of 30 W/m². Because the line conductors (made of aluminum alloys) are not insulated and can fall on the ground, the ROW strip should have no permanently inhabited structures and no tall vegetation (the limit is usually 1.8 m, high enough to grow Christmas trees under high-voltage lines).

Lines running through forests on flat or gently undulating land need adequate clearing, but many long spans that cross mountain valleys high above the ground do not require cleared strips. And when lines cross barren terrain, natural shrublands and grasslands, or land planted to field crops, there is also no need for ground clearances: existing land use can continue, and land claims, limited to the small areas needed to anchor transmission towers, can be further reduced with new monopole tower designs. Consequently, the high-voltage transmission land claims resemble those of wind turbines as they occupy only tiny shares (even less than 1%) of the total ROW claim. DC transmission has lower losses and narrower ROWs: to transmit the same power single-circuit 500-kV AC line would take 105 m, a double-circuit 500-kV AC line would claim 65–76 m, and 500-kV DC lines would need just 55–60 m of ROW (ATCO Electric 2010).

Manitoba Hydro's high-voltage DC link between the Nelson River power plants and converter stations near Winnipeg was a pioneering application (Manitoba Hydro 2013b). Its first phase, completed in 1973, has two bipole lines (900 kV, 895 km, and 1,000 kV, 937 km) and a capacity of about 4 GW. Two lines of 4,103 towers (spaced 427–488 m apart) claim a ROW of 137 m (about 12,300 ha, of which 10,800 ha are cleared forest) and have a power density of just over 30 W/m². The power densities for some other major high-voltage DC lines (60-m ROW for a 500-kV link) are close to 50 W/m² for each of the three lines from the Three Gorges Dam to Shanghai, Changzhou, and Guangdong, each with a capacity of 3 GW and a length of, respectively, 1,060, 890, and 940 km (Kumar, Ma, and Gou 2006). Formerly the world's longest high-voltage DC line, the 1,700-km link from the Inga Dam on the Congo to Kolwezi had an original design capacity of only 560 MW, but with a wide clearance of about 100 m its peak transmission power density is just 3 W/m^2 (Clerici 2007). The Xiangjiaba-Shanghai ultra-high-voltage 800-kV line, now the world's longest DC link, completed in 2010, spans about 2,000 km (ABB 2010).

The only transmission ROWs that can be attributed to a specific station are those of the lines built to connect it to the existing grid or dedicated high-voltage DC links from large hydro stations to distant cities. The only logical way to calculate the power densities of transmission is to quantify them on a national scale, that is, to divide the nationwide electricity generation by aggregate ROWs. And even that is not a correct solution for small, strongly interconnected countries engaged in vigorous electricity trade. For example, in 2011 Switzerland imported 83 TWh from the EU and exported 81 TWh, while its domestic consumption was only 59 TWh (Pauli 2013). I will present approximate calculations of global and US land claims in the penultimate chapter.

6 Energy Uses

Examinations of power densities of energy uses on scales ranging from global to local reveal quite a few unexpected (or at least unappreciated) realities. More important, historical perspectives shed light on one of the most important trends in the evolution of human societies, their quest for ever-higher power densities of energy use, be it on a collective level or as individuals.

In the preface I noted how land, a key concern of classical economics, became a marginal consideration in modern economies driven by concentrated labor and capital deployed in mass production. From a physically fundamental—that is, thermodynamic—perspective, this new pattern of economic organization is nothing but an expression of rising energy use, of deploying new energy conversions on unprecedented scales and with unprecedented intensities. This has resulted in a still continuing upward trend in the typical power densities of energy uses. This trend is an unmistakable but curiously unappreciated reality. Energy publications teem with data, with comparisons and analyses of energy use per capita, per unit of GDP, per unit of capital investment, more recently even per unit of energy (net energy return, or EROI, energy return on investment).

But energy use per unit of land is rarely investigated, and if so then in its most readily calculated, yet also most misleading, form, as annual energy use prorated over a nation's territory. This quotient may be relevant for Monaco, but its utility breaks down even at Hong Kong's level, as even this circumscribed and densely populated (de facto) city-state contains relatively large areas of steep (and largely deforested) mountain slopes (about two-thirds of its territory) frequented only by hikers and of swampland hosting waterbirds. And calculating such rates for dozens of countries where virtually all the population and all agricultural and industrial activities are concentrated on a small share (less than 10%) of the national territory is an exercise in deliberately introducing errors of one or more orders of magnitude. There are dozens of such nations. The largest ones include most of the countries of North Africa and the Middle East, but also Norway and Finland or Turkmenistan and Mongolia.

I note these misleading rates in passing as I follow power densities of energy uses from the planetary scale to microprocessors, the smallest massively deployed energy converters in modern civilization. But first, two important clarifications are in order, one definitional, the other one stressing the ever-changing, dynamic nature of energy use. This chapter is called "energy uses" rather than "energy consumption": is this just an idiosyncratic preference? Before I embarked on detailed reviews of the power densities of energy production I sorted out the key terms and provided their correct definitions. In energy studies this effort never ends, and I am not sure which of the pair of terms is misused more often: talking or writing about energy when what is meant is power or referring to energy consumption when what is meant is energy use (I too have been using the term energy consumption).

According to the first law of thermodynamics, that is an impossible feat: energy cannot be consumed, it remains conserved. It can be converted to a different form, and all of the conversions eventually end in dissipated heat that provides the feeble thermal backdrop of the known universe. This is not any verbal puritanism, for this critical distinction has important implications for energy use in modern societies: too often people think about energy as if it has been truly consumed, never again to be of any use or consequence. There are two important reasons why that belief is wrong. In the first place, even when we think that we are done with a particular energy conversion (that is, once we conclude that we have derived all useful work in the form of chemical, thermal, electric, kinetic, or electromagnetic energy), those energies have to be dissipated. The power densities of all of our final energy uses are also the power densities of heat rejection. Depending on the intensity and the scale of this heat rejection and on the heat-absorbing medium, such processes can cause significant temperature increases.

At one end of this heat-release spectrum are microprocessors, whose extraordinarily high power densities pose tremendous challenges for efficient heat dissipation. Among the most obvious objects much higher up on that size spectrum are giant cooling towers, the recipients of concentrated quantities of waste heat from the operation of large thermal electricity-generating plants. At the opposite end of the spectrum are modern megacities and conurbations. Specific heat rejections (per person, per vehicle) within these areas may be small, but their combined effect helps to create permanent urban heat islands that result in discernible changes in comfort, wind speed, and even precipitation.

In the second place, there are still too many instances in which the residual heat should not be left to dissipate, where it is unnecessarily wasted because we do not try hard enough to use it. I have already noted how combined-cycle electricity generation uses hot gas discharged by a gas turbine to vaporize water in a heat recovery steam generator and power an attached steam turbine to raise the overall efficiency of fuel use to 60%, roughly 50% above the usual performance of stand-alone gas turbines. And now the most common examples of harnessing "used" energy are hybrid vehicles, whose regenerative braking can recover some of the energy that is normally wasted in slowing a vehicle down by engaging an on-board electric motor as a temporary electricity generator that feeds a storage battery.

These examples of improving energy conversion efficiencies are also excellent illustrations of the second point I want to stress, the constantly changing levels of energy use. Unlike somatic energy requirements—with nutritional minima and optima delimited by human metabolism, which is itself a function of age, sex, and physical activity—there are no minima or optima for the use of extrasomatic energies. Their harnessing began with the use of draft animals. Preindustrial societies added conversions of water, wind, and biofuels, while affluent modern economies are expressions of large, incessant flows of fossil fuels and primary electricity.

But at every stage of this evolution there have been enormous differences in the level of use among countries and regions, with average per capita rates rising despite many impressive gains in conversion efficiencies. The quantifications of power densities of energy use in this chapter illustrate these differences and changes, but I am not trying either to forecast the future levels of energy use or to suggest optimal rates: the potential for higher conversion efficiencies—or for what I prefer to call a more rational use of energy (which may include avoiding certain conversions altogether)—and hence for reduced power densities of energy use remains large and ubiquitous, but how much of it will be realized will be determined by a complex interplay of social, economic, technical, and environmental factors.

I will pay attention to both aspects of energy use, as the final, controlled, deliberate conversions of fuels and electricity deployed to produce heat, motion, and light and as unavoidable processes dissipating heat into the environment. This means that I will appraise power densities on a descending scale from the planetary level to megacities and transport corridors to some major industrial process and individual buildings, all the way down to the now numerous indoor energy converters, and that I will also look at the power densities of heat rejection whenever they reach levels that either pose undesirable environmental problems or present great design challenges.

A Brief Historical Perspective

I open this segment with a brief look at the power densities of human metabolism. No energy conversion is obviously more important for our survival than food digestion, and as the evolution of human societies led first to higher densities of sedentary farming populations and then to even higher concentrations of anthropomass in cities, the power densities of human metabolism moved from negligible to substantial, and in many places still rising, values. The calculation of population-wide metabolic rates is a complex task that must take into account the age and sex structure of the societies studied and then apply appropriate activity adjustments to age- and sex-specific basal metabolic rates (Smil 2008). But to get fairly representative rates, it is much easier just to use data from metabolic models (Hall 2009) or from food intake surveys, which indicate that in affluent countries, the daily per capita means of food intake are mostly between 8.3 and 10 MJ (2,000–2,400 kcal/day).

In premodern societies these rates were considerably higher for adults because greater physical exertions (at least 15 MJ/day) were necessary to secure enough food and to energize the mining, transport, construction, and artisanal manufacturing that provided shelter and some material comforts. On the other hand, the smaller body sizes of most preindustrial populations and the common use of child labor tended to reduce average population-wide food requirements. Most people in premodern societies had barely adequate diets and owned little beyond often inadequate clothing and a small number of indispensable household items (Smil 2013b).

The population densities of foragers (hunters and gatherers) were as low as 1 person/10 km^2 of exploited land in marginal (arid, Arctic) environments and as high as 1/km^2 in coastal ecosystems where most food came from the ocean. The latter rate translates, even with heavy exertions in fishing and boat building, to a vanishingly low metabolic power density of 0.1 mW/m^2. Shifting cultivation raised the population densities to 20–30/km^2 (up to 4 mW/m^2), and traditional farming easily tripled or even quintupled those values, going from 100 people/km^2 of arable land in dynastic Egypt to 150 people/km^2 in medieval England and 400 people/km^2 (a metabolic power density of up to 45 mW/m^2) in intensively cultivated China in the late nineteenth century (Smil 2013a; fig. 6.1).

Even at the outset of the early modern era, populations were overwhelmingly rural: in 1500, cities contained perhaps no more than 4% of all people, and just 5% a century later (Klein Goldewijk, Beusen, and Janssen 2010). The largest cities of the ancient and medieval world had high population densities within their often massive protective walls. In 300 CE, imperial Rome housed about one million people in an area of just 15 km^2 enclosed by the Aurelian walls, the population density of nearly 67,000 people/km^2 (Smil 2010c) implying a metabolic power density of 7 W/m^2, a rate comparable to that of modern capitals.

The combination of post-1850 industrialization and rapid post-1950 population growth resulted in extensive urbanization: by 1900 15% of the global population lived in cities, by the year 2000 the share was about 47%. Using historical estimates of the total built-up area occupied by cities—10,000 km^2 in 1500, 47,000 km^2 in 1900, and 538,000 km^2 in the year 2000 (Klein Goldewijk, Beusen, and Janssen 2010)—results in the rise of average worldwide urban metabolic power densities from about 0.2 W/m^2 in 1500 to 0.5 W/m^2 in 1900 and to almost 0.6 W/m^2 by the year 2000. The urban share of the global population reached 50% in 2007; by the middle of the twenty-first century it is expected to approach 70% (UN 2012).

Extrasomatic Energies

In preindustrial societies these energies came overwhelmingly either from working animals or from the combustion of biofuels. My reconstruction of

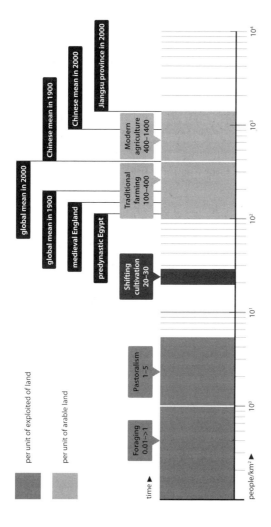

Figure 6.1
Evolution of population densities. Carl De Torres Graphic Design.

energy use in imperial Rome around 300 CE came up with about 100 MW of food energy (human metabolic power), less than 5 MW of feed energy (to sustain donkeys and horses engaged in urban transport), and at least 300 MW of wood and charcoal for cooking requirements, heating (including numerous baths with hot water pools), and artisanal manufacturing (Smil 2010c). This prorates to a power density of 25 W/m^2 within the Aurelian walls, a high rate reflecting the high population density during the later imperial era and the low efficiency of biofuel combustion.

Not surprisingly, as long as cities remained densely populated (confined by walls), and as long as animate energies and the combustion of biofuels remained the only energy inputs, the highest power densities of urban energy use remained very similar. Galloway, Keene, and Murphy (1996) put London's wood consumption at the beginning of the fourteenth century at about 25 GJ/capita, implying an annual power of about 63 MW for the town's 80,000 inhabitants; adding their metabolism and the (relatively unimportant) feed for horses raises the total to less than 75 MW, and prorated over the area of some 1.8 km^2 that yields an overall power density of London's energy use in 1300 of roughly 40 W/m^2.

Initially, industrialization brought a counterintuitive shift in the power densities of energy use: in large European cities they began to decline during the early and middle decades of the nineteenth century even as the inputs of coal (for households, industries, and transportation) were rising, and as (starting in the early 1880s) coal-fired power plants (all originally situated in cities) created a new large fuel demand. This is easily explained by the rapid spatial growth of modern cities. For example, London's area expanded more than tenfold during the nineteenth century, from about 24 to 280 km^2 (Demographia 2014). Paris expanded at a slower rate, from 34 km^2 in the early eighteenth century to just over 100 km^2 two centuries later (with its population growing from 600,000 in 1700 to 2.7 million in 1900). A study by Kim and Barles (2012) makes it possible to follow the city's rising, and changing, energy demand.

By 1800 the city's average per capita demand for all extrasomatic energies was almost 30 GJ/capita, with wood dominant; coal began to supply more than half of the city's energy demand by 1860, and even as the population expanded nearly fivefold during the nineteenth century, the average per capita energy use remained at about 25 GJ, reflecting the higher efficiencies of modern conversions. These shifts translate to almost 500 MW of wood and charcoal in 1800 and 2.1 GW (mostly coal) in 1900, and pro-rate to power densities of nearly 15 and 20 W/m^2. The addition of meta-bolic energies would raise these rates only slightly, to, respectively, 16 and 23 W/m^2, as expanding fuel use relegated somatic energies to totals lower than the inherent errors in estimating extrasomatic inputs.

Fuel, and later also electricity, requirements for the most energy-intensive and spatially concentrated industrial processes brought unprecedented power densities, as exemplified by contrasting one of the most advanced large-scale manufacturing enterprises of the early nineteenth century with the signature facility of the most important branch of early twentieth-century manufacturing. The Merrimack Manufacturing Company, Ameri-ca's first large and fully integrated clothing producer (mainly of calico fabrics), opened in 1823 in Lowell, Massachusetts (Malone 2009). The plant occupied about 10 ha and drew about 2 MW of water power from a large (10-m) drop in the Merrimack River. The plant operated with a power density of about 20 W/m^2 when prorated over its entire area, and about 50 W/m^2 when prorated over its actual floor space.

Almost exactly a century later, in 1928, Henry Ford's River Rouge plant began to produce cars on about 3.6 km^2 of land west of Detroit in a com-pletely integrated complex where all essential inputs (coke, steel, glass) and components (steel plates, forgings, engines) were made on-site. During its peak production the plant employed more than 100,000 people and pro-duced a car every 50 seconds (The Henry Ford 2013; fig. 6.2). The primary energy inputs at the River Rouge plant—dominated by coal to make about 1.5 Mt coke/year and to generate electricity in a plant with an installed capacity of 375 MW$_e$—amounted to nearly 3.5 GW, resulting in a power density in excess of 1,000 W/m^2 when prorated over the entire site and about 2,500 W/m^2 when 1.42 Mm^2 of floor space is used as the denomina-tor. Consequently, the two great phases of modern industrialization, set a century apart, had roughly a 50-fold difference in their operating power density.

Operating power densities on the order of 10^3 W/m^2 became common in large metallurgical enterprises as well as in new large refineries producing a wider range of liquid fuels and gases with higher efficiency. Pennsylvania's Homestead Steel Works, bought in 1883 by Andrew Carnegie and expanded to become the largest component of his eponymous steel company, are among the best illustrations of these high operating power densities. The

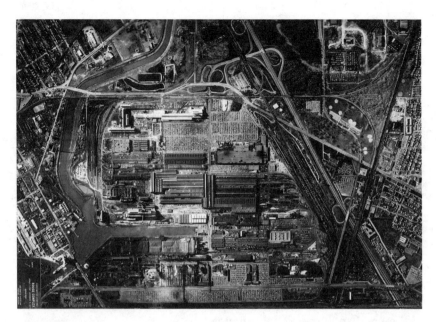

Figure 6.2
Ford Motor Company's River Rouge plant, aerial view. Ford Motor History (n.d.).

complex required about 300 PJ of coke, coal, and natural gas for some 6 Mt of finished steel products (Carnegie Steel Company 2012). The works occupied about a 112-ha site on the southern bank of the Monongahela east of Pittsburgh, and their operating power density of about 2,000 W/m² remained characteristic of large iron and steel mills for most of the twentieth century.

Finally, I must stress that the power densities of agricultural operations have seen increases far higher than the rates in those industrial processes that moved from small-scale artisanal operations to large-scale high-throughput enterprises. The best example is the cultivation of grain crops used for food and feed. Traditional low-yield grain farming (wheat or rice yields no higher than 1.5 t/ha) was based solely on animate (human and draft animal) labor and on recycling organic wastes and rotations with leguminous crops (Smil 2008). Detailed accounts by Buck (1930, 1937) show that even in relatively high-yielding, irrigated fields in China, the rice harvest required the deployment of less than 7 GJ of animate metabolic energies, more than 95% of it in the form of the draft power of water buffaloes.

When the power density of this traditional cropping is calculated by using actual time worked in the fields (rather than the cropping period of about 150 days or, as in other power density calculations in this book, the entire year), human and animal exertion were deployed at rates lower than 1 W/m² of arable land (Smil 2008).

In contrast, the direct energy investment in modern high-yield cropping is dominated by liquid fuels (diesel oil, gasoline) for machinery (used for plowing, planting, the application of agrochemicals, harvesting), and additional fuels or electricity are required for irrigation. Corn, America's principal feed crop, now takes only about seven hours of labor per hectare to produce high (10–11 t/ha) yields in Iowa (Duffy 2013). Typical diesel fuel requirements are 17–20 L/ha for plowing, 10–15 L/ha for disking, 5 L/ha for planting, a similar amount of fuel for fertilizer application, and 15 L/ha for combining (Grisso et al. 2010). The overall fuel requirement will be no less than 65 L/ha (2.3 GJ/ha), which means that the power density of direct energy uses will be close to 10 W/m² of cultivated land, an order of magnitude more than in traditional cropping.

Fuel requirements can easily double in heavier soils and with irrigation drawing water from deep aquifers, and indirect energy needs (above all those to synthesize fertilizers, pesticides, and insecticides) may double that larger total. Attributing these indirect needs would require a different definition of power densities of energy use, one that would have to

Box 6.1
Animate energies in traditional crop cultivation

Traditional rice cultivation in China: water buffalo, crop yield 1.4 t/ha

300 hours of human labor × 700 kJ/hour = 210 MJ/300 = ~200 W

250 hours of animal labor × 25 MJ/hour = 6.25 GJ/250 = ~7,000 W

7,200 W/10,000 m² = 0.72 W/m²

Traditional wheat cultivation in the Netherlands: two horses, crop yield 2 t/ha

170 hours of human labor × 700 kJ/hour = 120 MJ/170 = ~195 W

120 hours of animal labor × 25 MJ/hour = 3 GJ/120 = ~6,900 W

7,100 W/10,000 m² = 0.71 W/m²

consider (to give just one obvious example) not only fuels and electricity used by households but also the energy required to produce building materials and the energy embodied in furnishings, appliances, and other domestic items.

This brief retrospective makes several trends clear. The power densities of human metabolism rose by four, even five orders of magnitude as human societies advanced from small groups of foragers to inhabitants of large cities. Well into the early modern era, extrasomatic energies were dominated by the metabolic conversion of working animals and by biofuels. In cities, the rising use of wood and charcoal, and in a few countries also a higher reliance on coal, relegated all animate metabolism to a marginal role in energy supply. This trend became even more pronounced with advancing industrialization, which created numerous production sites (for metallurgy, refineries, chemical syntheses, intensive manufacturing) where energy uses proceed with power densities of 10^3 W/m^2. The next section takes a revealing look at modern power densities of aggregate energy use by descending the spatial scale.

Hierarchy of Modern Energy Uses

I will move from global to local, presenting first large-scale power densities merely as interesting rates—some, as already noted, misleading, others actually quite revealing—before descending to scales that really matter, to the power densities of modern cities, some key industries, transportation corridors, and individual buildings. In most of these instances heat rejection is just a different label for the process of energy use: the process simply takes place, and it requires no specific attention; but in many instances the power densities involved in getting rid of heat are surprisingly high, in some cases reaching truly astonishing rates, and that is why I single them out for a closer look in the next segment of this chapter.

Global Scale

The worldwide power density of human energy use is a perfect example of a misleading quantification. The Earth, with a mean radius of 6,371 km, has a surface area of 510 Tm^2, with oceans covering 361 Tm^2, dry land 149 Tm^2, and ice-free land about 133 Tm^2. With global energy use rising from 1.38 TW in 1900 to 12.43 TW in 2000, this means that on the planetary level, the power density of primary energy use rose during the twentieth century by almost exactly an order of magnitude, from 0.0027 to 0.024 W/m^2. The first 13 years of the twenty-first century saw (mainly thanks to Asia in general, and China in particular) a further rapid rise, to almost exactly 17 TW, and hence the power density of global energy use prorating to 0.033 W/m^2 of the Earth's surface. But this rate can be actually observed at very few places on Earth as the distribution of global energy use remains highly uneven.

Recent totals of global population, extent of economic activity, and density of transportation links are all unprecedented, but vast areas of the ocean and large chunks of continental masses remain places where no continuous conversions of anthropogenic energy is under way, or where highly intermittent energy uses consist of a lone ship traversing some extreme latitudes or a jetliner on one of the least frequented routes. The simplest correction to move the rate closer to its mean or modal value is to prorate the energy use over ice-free land, where most fuels and electricity are produced and used. This quadruples the global 2013 power density, to 0.125 W/m^2, but the flux is still only a negligible fraction of the mean insolation received by the continents, just 0.066% of 188 W/m^2.

Moreover, the total anthropogenic radiative forcing, largely resulting from the emissions of greenhouse gases, has already reached 2.3 (1.1–3.3) W/m^2 since the beginning of the industrial era (IPCC 2013). Consequently, even another doubling of global energy use that would raise the mean continental heat dissipation to 0.25 W/m^2 would still remain an order of magnitude below the current radiative impact of greenhouse gases; moreover, it would also remain much lower than the margin of error associated with quantifying such uncertain components of the Earth's radiation balance as the tropospheric ozone and cloud albedo effect resulting from the presence of airborne aerosols.

In any case, prorating global energy solely over the continental area is an unsatisfactory correction as it ignores two obvious realities: parts of ice-free land have only a negligible, or exceedingly patchy, population presence, and fuels are also converted by shipping and intercontinental flights. Parts of the ocean are traversed by relatively frequent and regular shipping routes (especially oil tanker routes from the Middle East to East Asia and Europe, and routes followed by container vessels from East Asia to North America and Europe) and are overflown by jetliners (particularly the northern Atlantic and the northern Pacific Ocean). The best choice would be to calculate two rates, one with the denominator encompassing all of the relatively densely populated regions on land, the other one with the denominator aggregating shipping and air routes across the ocean. Only the first adjustment can be done with satisfactory accuracy.

A more realistic expression of the power density of continental energy use is to restrict the denominator to urban and industrial areas and their transportation corridors. Ten global assessments of urban (or urban-related) areas published between 1992 and 2009 and reviewed by Schneider, Friedl, and Potere (2009) resulted in aggregates ranging over an order of magnitude, with the lowest estimate at just 276,000 km^2 (for areas defined as populated places based on digitized maps) to as much as 3.52 Tm^2 for urban extent based on a combination of census data, maps, and nighttime satellite images. A critical assessment of these studies by Potere and colleagues (2009)—based mainly on a random sample of 10,000 validation sites analyzed in high resolution—found that Schneider, Friedl, and Potere (2009) offered the most accurate result, with 657,000 km^2 of land where built structures covered more than 50% of the surface in the year 2001. Even if that total grew to about 800,000 km^2 by 2013, it would still be just 0.6% of ice-free land.

The first-order approximation would be to assume that at least 70% of all economic activity (and hence of all energy use) takes place in that relatively small area, which would result in a global power density of about 19 W/m^2 for the year 2000 and 21 W/m^2 for 2012. But using urban land is an imperfect correction as cities are quite heterogeneous: they include not only a large share of impervious surface areas (ISA, including all roofed and paved surfaces devoid of vegetation) but also the variable, and often extensive, grass and tree cover in residential districts, along city streets, highways, and railroads, and in parks. The best way to estimate the power densities of energy use on a global or national scale is to use the aggregates of ISA whose extent can be fairly satisfactorily estimated from satellite imagery. Elvidge and co-workers (2007) pieced together the first global account of ISA for the years 2000–2001. Their study found a global ISA total of about 580,000 km^2, that is, just 0.43% of ice-free land, and an aggregate equivalent of Kenya or Madagascar.

Adjusting the numerator can be done in two ways: by subtracting energy use outside urban energy use, and, in addition, by subtracting all transportation energies taken onboard in cities but used for intercity traffic. The first adjustment can be done by calculating the weighted mean based on the continental shares of urban populations. In Europe, the Americas, and Australia, as well as in Japan and South Korea, urban populations account for 75%–85% of the total population and (because their per capita energy use is higher than a national average) use about 85% of all energy; in Africa and in Asia, urban populations account for about 45%–50% of the total and use about 55% of all energy. The weighted global mean is thus about 75% of energy used in urban or industrial regions, and that total (12.5 TW in 2012), prorated over 580,000 km^2, results in a power density of about 22 W/m^2.

In the second case we have to subtract all energy for intercity shipping, aviation, and road and rail traffic. The global fuel demand for oceangoing vessels has recently been about 350 Mtoe/year and for aviation about 250 Mtoe/year. If we assume that 35%–40% of all road traffic is outside urban areas, that leaves us with about 40% of all transportation fuel (roughly 1.25 TW out of the total of 3.2 TW) used in urban regions. The aggregate urban energy use is thus reduced to 40% of transportation energy (roughly 1.3 TW) and 75% of all other uses (or 10.1 TW) and, with an urban demand of 11.4 TW, the average power density of urban energy uses is just below 20 W/m^2.

Because of the great land cover and land-use disparities and the large economic differences between areas, continental averages of power densities are hardly more revealing than global means, and as long as we use entire national territories in numerators we do not get closer to truly informative rates even once we step down to the state level. The extremes of these unrepresentative power densities range from negligible rates for nearly all Sahelian countries, which generally have the world's lowest per capita energy use and large desert or arid grassland territories. When USEIA data for 2010 are used the rates range from 0.0003 W/m^2 for Mali and no more than 0.0005 W/m^2 for Niger to values four orders of magnitude higher for small, densely populated nations with affluent, high-energy economies: the Netherlands at about 3.4 W/m^2, South Korea at 3.6 W/m^2.

Singapore, with a power density of nearly 140 W/m^2, rates significantly higher, but the power density of this city-state is particularly misleading because most of the energy purchased and processed in the city is not for its own use but is fuel exported by its giant refineries and the fuel oil, diesel, and kerosene taken onboard ships and jetliners in one of the world's most important shipping and air travel hubs. Among larger, more populous nations, Japan (1.9 W/m²) and Germany (1.2 W/m²) stand out because of their relatively high power densities. In contrast, the mean for China, although it quadrupled between 1978 and 2008, is only 0.35 W/m². And Canada's high per capita energy use cannot compensate for the country's large territory, and so the nationwide power density of energy use is just 0.045 W/m², an order of magnitude below the US rate (0.33 W/m², including Alaska and Hawaii).

Urban Power Densities

The power densities of energy use become meaningful only when the annual supply of fuels and electricity is prorated over those areas where most of it gets converted to final uses (or at least taken onboard), that is, when adjustments are made for energy use in a nation's urban areas. We now have at least four studies of the nationwide extent of urban areas or the ISA in the United States. The US Geological Survey put the first nationwide estimate of urban areas at 90,000 km² (USGS 2000). The study that combined nighttime lights' radiance with LANDSAT imagery put the total extent of the US ISA at nearly 113,000 ± 13,000 km², slightly less than the surface of Ohio (Elvidge et al. 2004). Elvidge and co-workers (2007) lowered the total to about 84,000 km², close to the USGS study figures. Churkina, Brown, and Keoleian (2010) came up with a higher total of 141,000 ± 40,000 km² for the year 2000, and they also calculated that grassy surfaces covered about 40% and treed surfaces extended over about 25% of America's urban areas.

Correcting for these vegetated surfaces brings the mean total of the true US ISA to about 50,000 km². For final energy uses I assume first that 80% of electricity is generated outside urban areas, and then I assume that 85% of all residential, 90% of all commercial, 75% of all industrial, and 50% of all transportation energy use take place in urban areas; their annual energy demand would have amounted to about 55% of the countrywide total of 3.25 TW in 2011, or nearly 1.8 TW. With the extremes of 84,000 km² and 180,000 km² in the denominator, this would translate to 10–21 W/m². Excluding all vegetated surface and putting 50,000 km² in the denominator raises the average US urban power density of energy conversions to about

35 W/m². Power densities of 10–35 W/m² are well supported by rates calculated by bottom-up sectoral aggregations for specific US cities, as well as for urban areas in an increasing number of other (mainly European and Asian) countries.

Recent work on anthropogenic heat releases spans scales from global assessments to appraisals at national and city level and to microstudies of specific wards in large cities (Allen, Lindberg, and Grimmond 2010; Chen et al. 2012; Hsieh, Aramaki, and Hanaki 2007; Wong, Dai, and Paul 2012). There are also many studies of urban heat islands. Peng and colleagues (2011) quantified them for 419 of the world's big cities, but their results cannot be readily translated into the power density of specific urban energy use. And I should note that there is also a downward anthropogenic heat flux creating a subsurface urban heat island, but, naturally, its intensity is a tiny fraction of the surface counterpart. Menberg and co-workers (2013) actually calculated that in Karlsruhe, the average total heat flux into the city's shallow aquifer was about 760 ± 89 mW/m² in 1977 and that by 2011 it had increased to nearly 830 ± 143 mW/m².

The power densities of urban energy dissipation have several common attributes. They display expected daily, weekly, and seasonal fluctuations (daily maxima between 11 AM and 6 PM local time; weekend lows; winter maxima in cold regions). Their annual means range between 10 and 100 W/m², but the seasonal extremes for the smaller areas with the highest energy use go up to, and even above, 1,000 W/m², exceeding the energy received in some locations even during the noontime peak. In the past, cities in cold climates had the highest pronounced winter peaks, but very high rates now prevail even in tropical climates with the emergence of high-rise-filled downtowns and ubiquitous air conditioning.

Bottom-up approaches used to quantify the power densities of urban energy use rely on a wide range of relevant statistics (population densities, human metabolism, energy use by buildings and industries, the density of road transport, the specific energy demand of vehicles) incorporated into often detailed models. Quah and Roth (2012) presented more than two scores of annual (and also seasonal) power densities published between 1952 and 2009 for cities in the United States, Canada, Europe, and Asia. The annual extremes in this set ranged from just 4–5 W/m² for suburban areas of smaller cities (Swiss Basel and Polish Łódź) to 159 W/m² for Manhattan in 1967. The modal range was between 11 and 30 W/m², and Tokyo in 1989 had the hourly extremes (for both summer and winter) at, respectively, 908 W/m^2 and 1,590 W/m^2 (Ichinose, Shimodozono, and Hanaki 1999). Among notable recent bottom-up calculations are those of Hsieh, Aramaki, and Hanaki (2007) for Taipei; Iamarino, Beevers, and Grimmond (2012) for London; Quah and Roth (2012) for Singapore; and Howard and co-workers (2012) for New York City.

A high-resolution (both in space and time) assessment for London shows that human metabolism contributed just 5% of the mean annual anthropogenic flux of 10.8 W/m^2 for 2005–2008, that domestic and industrial sources were nearly equal (4.6 and 4.1 W/m^2), and that road fuels contributed about 15% of the total (1.65 W/m^2). As expected, the rates declined toward the outskirts of Greater London, and less than 3% of the city had values above 50 W/m^2, with the City of London (highest density of high-rise buildings, 348,000 daily workers within 3.2 km^2 in 1007) going up to 140 W/m^2, with local peaks up to 210 W/m^2. Sensible heat dominated the flux, latent heat carried away only about 7% of thermal energy, and heat transferred first to wastewater accounted for about 12% (Iamarino, Beevers, and Grimmond 2012).

In Singapore, 24-hour maxima reached 113 W/m^2 in the commercial district, 17 W/m^2 in high-density public housing, and 13 W/m^2 in the low-density residential areas, with buildings (primarily due to cooling) being the largest source of dissipated heat, 49%–82% on weekdays and 46%–81% on weekends (Quah and Roth 2012). In Da-an ward of Taipei city, the daily heat rejection by air-conditioned buildings averaged 15 W/m^2 for residential housing and 75 W/m^2 for commercial establishments, with an overall mean of 34 W/m^2 (Hsieh, Aramaki, and Hanaki 2007). The most informative of these studies of urban energy use is a citywide mapping on a block and lot level for New York (Howard et al. 2012).

By far the most interesting part of that study is a map of power densities using block land area in the denominator. Some mid-Manhattan blocks with a high density of high-rises show power densities as high as 900 W/m^2, which means that a single block demands a year-round average of nearly 18 MW of delivered energy. Other parts of the city where many blocks have power densities in excess of 400 or 500 W/m^2 are the financial district, Greenwich Village, the Flatiron, Midtown South, Sutton Place, and the East Side. Manhattan's lowest power densities are in Harlem and East and West Village, and densities in residential boroughs are generally below 25 W/m^2, while city blocks in parts of Queens and on Staten Island rate less than 15 W/m^2.

Hong Kong's energy statistics list end uses for specific sectors and detailed land-use data make it easy to reconstruct power densities on an annual basis (Government of Hong Kong 2013). In 2012 the average power density of residential energy use was about 40 W/m^2, a high rate reflecting the dominance of the city's crowded high-rise housing estates; the power density of industrial areas was about 20 W/m^2, and that of land transportation reached a very high rate of about 50 W/m^2, while the overall rate for the city's impervious surfaces (including a large and busy port and airport) was roughly 60 W/m^2.

Power Densities of Buildings

The order of magnitude does not change as we narrow the spatial focus from the most densely built-up city wards to individual buildings or from areas of heaviest traffic to individual roads or crossroads. Nationwide means of power densities of America's building stock can be calculated from statistics published in the *Buildings Energy Data Book*, a compilation of the USDOE (2013). Its disaggregated annual totals show the 2010 site use average (that is, with fuel-based electricity not converted to its primary energy equivalent) of 40 W/m^2 for commercial buildings (with space heating claiming 27%, cooling 10%, and water heating less than 7%) and just over 16 W/m^2 for residential buildings, with about 45% of that rate due to space heating, more than 16% used by water heating, and about 9% by cooling.

US residential lighting needs less than 6% of all electricity, with a power density of just below 1 W/m^2. Commercial indoor lighting claims almost 14% of the total usage (about 5.5 W/m^2). Disaggregations by building function for New York City by Howard and co-workers (2012) showed energy use per floor area ranging from just over 20 W/m^2 for residential housing (for one to four families) to about 35 W/m^2 for multifamily residential housing and almost 70 W/m^2 for stores, schools, and hospitals. One of America's best sources for power densities of energy use in a large number of specific buildings is San Francisco's annual *Energy Benchmarking Report*, which lists the electricity and fuel uses of hundreds of municipal structures (San Francisco Water Power Sewer 2013).

Modal ranges in the ascending order of densities for 2012 are (all values in W/m^2 of floor area and for energy used on-site): parking garages,

2–6 W/m²; warehouses, 5–11 W/m²; schools, 15–22 W/m²; offices, libraries, performance halls, conventions centers, and police and fire stations, 22–35 W/m²; and museums, 35–100 W/m². As expected, modern offices, schools, and retail space have roughly similar specific energy requirements, on the order of 20–30 W/m² of floor area, less than half as much as supermarkets (they were even more wasteful in the past due to open-bin freezers) and as little as one-third of the rate for busy restaurants and hospitals.

I should also note that some studies express demand in terms of primary energy, roughly doubling the usually quoted end-use values. That is, of course, an inevitable consequence of two realities: all modern buildings require large inputs of electricity for lighting, air conditioning, appliances, and electronic devices (some also for heating and water heating), and in most countries (including the United States), most of the delivered electricity comes from thermal generation, with its inherently large conversion losses (the best efficiencies are now around 40%). But, as applied in this book, the power densities of energy use measure only actual on-site consumption.

But another adjustment is necessary. As I have clearly indicated, the published densities of energy use in buildings refer to requirements per unit of floor space, and hence they equal the power density as defined in this book only for single-story buildings and so must be appropriately enlarged for multistory structures. In any case, published data must make it clear whether the rates refer to a unit of floor area or to a structure's footprint. And comparing the energy performance of buildings simply by referring to their power densities is misleading if such a comparison is intended to illustrate differences in efficiency. Corrections must be made for the average number of heating and cooling days, for the primary sources of used energy, and for the ownership of electricity-consuming appliances and electronic devices; moreover, preferred indoor temperatures and appropriate levels of lighting should also be taken into account.

For example, my superinsulated two-story house (2″ × 6″ frame with fiberglass in the walls, a thick layer of blown insulation in the attic, an exterior Styrofoam wrap around the foundation, triple windows with argon, a 97% efficient natural gas furnace, heat recovery ventilator) will use more energy than an indifferently built house in Vancouver. The two cities have the same latitude (50°N), but during 2013 Winnipeg's mid-continental location had 2.6 times more heating-degree days than Vancouver's much warmer maritime climate (Degree Days 2014). Such differences will always remain, but the historical evidence is clear: better construction, more efficient heating and cooling, fewer electricity-intensive appliances, and less wasteful lighting have brought impressive declines in the average residential power density.

Many fairly large differences in the power densities of individual buildings are more due to design and operating practices than to disparities in climates and preferred indoor temperatures. European data show that the power densities of single-family house heating in Germany hardly changed between 1918 and 1957 (nearly 30 W/m^2 of floor area), but then declined to 18 W/m^2 by 1967 and to just 6 W/m^2 in 2010. In the UK the drop was from about 66 W/m^2 of floor area before 1920 to 23 W/m^2 by 2002, and in Italy from 25 W/m^2 to about 10 W/m^2 between 1950 and 2005 (BPIE 2011). Between 1980 and 2005 the average US decline was 37% for all housing units and 45% for single-family detached houses in the South (USEIA 2009).

New office buildings are also much more energy-efficient. For example, Canada's most efficient office building, the headquarters of Enermodal Engineering, in Kitchener, Ontario, draws only 8.5 W/m^2 of floor area, only a tenth of the country's typical multistory office structure (Enermodal Engineering 2013). The most energy-efficient building in the United States, Seattle's Bullitt Center (with a large PV array, geothermal heating and cooling, and motorized windows), is in a much warmer climate; its demand comes at just 6 W/m^2 of floor area (Bullitt Center 2013). The best house designs can result in similarly low rates, but a zero-energy house in any colder climate is a misnomer: the house may not need any external source of energy, but even with an efficient passive design it will require a substantial investment in a rooftop PV array or a geothermal system, or both.

Finally, it is revealing to add up energy uses per unit of floor area to get the ascending range of power densities of common, as well exceptional, buildings. An energy-efficient single-story house in a mild maritime climate that requires minimum heating and no cooling could average less than 10 W/m^2. The two just described superefficient office buildings are both low-rises: Enermodal is only three stories high and the Bullitt Center has just six stories, and hence their average power densities are respectively just 25.5 and 48 W/m^2. A modern (post-1990) American detached two-story

Figure 6.3
Kwai Chung housing, Hong Kong. © JEROME FAVRE/epa/Corbis.

house will average between 30 and 40 W/m^2 of its foundation, and an older (pre-1980) 20-story office building in a climate that requires both heating and cooling will average 800 W/m^2. Buildings in Kwai Chung, Hong Kong's largest public housing estate, in New Territories, with 16 towers of 38 floors and almost 25 W/m^2 of floor area, will have a power density of roughly 950 W/m^2; that is (as already shown) the same as the most energy-intensive city blocks in Manhattan's Midtown (fig. 6.3).

Midprice and luxury hotels have above-average energy needs. The average for US hotels has been 45 W/m^2 per unit of floor area, whereas in much colder Ottawa it is 77 W/m^2, in London (because of poorly insulated buildings and inefficient heating) as much as 80 W/m^2, and in Hong Kong (because of air conditioning) 63 W/m^2, but the average in Auckland's moderate climate is only about 30 W/m^2 (Deng and Burnett 2000; Su 2012). This means that 10-story buildings will have power densities of up to 800 W/m^2 and that the tendency to raise many modern luxury hotels to more than 50 stories creates exceptionally high power densities, particularly in deserts. A 50-story hotel in a hot climate has a power density close to, or in excess of, 2,000 W/m^2, and Burj Khalifa, the world's tallest building

(828 m, 160 floors), in Dubai, has a base footprint of 8,000 m^2 and a peak electricity demand of 50 MW, implying short-term power densities of up to 6,250 W/m^2 of its foundation.

Transportation Densities

Urban traffic is usually a much smaller contributor to urban energy use than domestic demand and industrial processes. For example, disaggregated data for Greater London between 2005 and 2008 show the three shares at, respectively, 15%, 42% and 38%, but urban traffic can reach very high power densities along major heavily traveled roads even when it flows freely, and even higher rates in prolonged traffic jams when drivers keep their engines idling. This is an inevitable consequence of the low efficiency of gasoline engines: they lose about 30% of initial energy input through their radiators and about 40% in exhaust gases.

Most of the latter flux is redistributed by radiation before it exits a tail-pipe: exhaust temperature is less than 70°C, while the gases leaving the cylinder have a temperature of about 800°C. Both radiator heat and exhaust heat are absorbed first by other car structures before they are lost to the atmosphere. The logical way is thus to use a car's footprint rather than just the footprint of a radiator or exhaust system as the denominator in calculating vehicular power densities. Small cars have a footprint of less than 4 m^2 (3.8 m^2 for the Honda Fit), while large ones go over 5 m^2 (5.1 m^2 for a Mercedes-Benz S class). With power ratings of, respectively, about 87 kW and 339 kW, the maximum theoretical power densities of energy use by those two vehicles would be 22,900 W_t/m^2 and 66,500 W_t/m^2 of their footprint, rates resembling the power densities of heat dissipation in large power plant cooling towers.

Dissipated heat is diffused along roads, driveways, and parking lots. The maximum short-term rates of that dissipation depend on roadbed widths (3.6 m is the US standard for freeways, 2.7–3.6 m for local roads), speeds, distances between vehicles, and their power rating. Free-flowing car traffic on a freeway (30 cars/km per lane traveling at 80 km/hour) with cars averaging 8 L/100 km (29 mpg in the United States) will have a power density of about 475 W/m^2 of paved lane. A mixture of 70% cars and 30% trucks with 1,000 vehicles/km of a single lane in one hour would generate a power density of about 560 W/m^2:

Box 6.2

Power density of highway traffic

1,000 vehicles/hour

700 cars and 300 trucks, lane width 3.6 m

cars 4 kJ/m 700×4 kJ $\times 1,000$ m $= 2.8$ GJ

trucks 15 kJ/m 300×15 kJ $\times 1,000$ m $= 4.5$ GJ

7.3 GJ/3,600 = 2.03 MW/3,600 m2 = 563 W/m2

In a traffic jam, with 125 idling vehicles per kilometer of street lane just 3 m wide and consuming about 1.3 MJ/minute (roughly 21 kW per vehicle), the stationary power density will reach 900 W/m^2. These high rates apply only to the lanes. Including the associated road infrastructure in the calculations usually halves those rates and often reduces them even more. Paved shoulders on US interstate highways have minimum width of 3.05 m on the outside and 1.22 m on the inside, while medians have a minimum width of 11 m in rural and 3 m in urban areas. Including these surfaces for a four-lane highway would reduce all of the just calculated rates by at least 57%, and by more than 60% in rural settings, resulting in rates below 250 W/m^2 in flowing traffic. Including adjacent land that forms the entire road infrastructure (ditches, embankments, land cut off by approaches, exits, and interchanges) would bring further reductions on the order of 20%-40%, to rates well below 200 W/m^2 for the entire transportation corridor in flowing high-density traffic.

Adjusting these calculated short-term peak rates for average annual traffic, with its enormous fluctuations between nearly empty roads in the early morning hours and seemingly endless congestions during rush hours, cuts the rates by an order of magnitude. The next adjustment to make in calculating the long-term power density of urban traffic is to prorate it over the entire area devoted to roads and parking lots, with the former claiming about 30% of all land (35% in many US cities) and the latter at least 20% of the total. Prorating the road transport energy use over the entire city area will cut the density by about two-thirds, to rates below 5 W/m^2, compared to the rates derived by having only road infrastructure as the denominator.

The actual means of urban traffic power densities in Greater London during the years 2005–2008 were 1.65 W/m^2 (Iamarino, Beevers, and Grimmond 2012). Recent gasoline consumption data for Los Angeles County, the epitome of high-density traffic, show annual purchases running at a rate of about 72 GW. Even if it assumed that only one-third of all gasoline purchased in the county is actually consumed within its borders, this would prorate to 2 W/m^2 for the county's entire area, or 4 W/m^2 for its residential and industrial land, and adding diesel fuel (assuming the nationwide proportion of gasoline to diesel use is valid for Los Angeles) would raise this to at least 5.5 W/m^2.

The Challenges of Heat Rejection

Many energy conversions do not require any special arrangements to dissipate heat, just cautious handling of hot light bulbs, toasters, and irons; in many other instances simple arrangements, such as radiator fins, suffice. But even the most efficient large thermal electricity-generating stations reject about 60% of the initial energy input as waste heat, and such large fluxes require appropriate heat-disposal designs. As already noted, cooling towers reject about 50% of all energy used by thermal power plants, and large natural draft units (massive concrete structures up to 100 m in diameter and up to 200 m tall) can handle 20–40 kW$_i$/m^2. With a 40% conversion efficiency a 2-GW$_i$ plant will reject about 2.5 GW$_t$ through its cooling towers, and the actual heat rejection power densities will be close to 80,000 W$_t$/m^2 of the tower footprint. The overall land claim will be larger because the towers must be set well apart to prevent Venturi effect–induced wind loads on their relatively slender walls. Cross-flow cooling towers using mechanical draft units are even more compact, with throughput densities of 100,000–125,000 W$_t$/m^2.

About 10% of a plant's initial fuel input is rejected through a stack. At a 40% conversion efficiency, this amounts to about 500 MW$_t$ for a 2-GW$_i$ station. Tall chimneys (scores of the tallest ones have surpassed 300 m) have a top inside diameter of just 3–7 m and a bottom inside diameter of 14–17 m, which means that they reject heat with power densities of 0.7–1 MW$_t$/m^2 of their foundations. The introduction of flue-gas desulfurization changed that because flue gases (120–150°C) must be cooled to saturation temperature before their SO$_2$ reacts with alkaline compounds, and they leave stacks at less than 50°C, producing heat-rejection power densities an order of magnitude lower than they did previously.

The greatest technical heat rejection challenge takes place on a microscale, thanks to the ever-increasing crowding of microprocessor transistors (fig. 6.4). Intel 4004, the world's first microprocessor, released in 1971, had 2,300 transistors on a 135 mm^2 silicon die, and dissipated about 2.5 W$_t$/cm^2 (Intel 2013). In 1978 the Intel 8086 contained 29,000 transistors and dissipated 7.6 W$_t$/cm^2, almost exactly as much as a small (160 cm^2) circular kitchen hot plate rated at 1,200 W. By the beginning of the twenty-first century Intel's ultra-large-scale integration procedures had crowded 50 million transistors on 130 mm^2 of the Xeon Irwindale, whose demand of 115–130 W prorated to up to 100 W$_t$/cm^2, equivalent to 1 MW$_t$/m^2, and in 2005 the Pentium 4 went above 100 W$_t$/cm^2 (Azar 2000; Intel 2005; Joshi 2001).

Microchips are usually placed in very tight confinement, but to ensure optimum performance, their temperature should be kept below 45°C or else their operation will slow down and eventually cease. The challenge is particularly great when dealing with hot spots, whose fluxes can briefly go as high 1,000 W$_t$/cm^2; this power density is equal to about 15% of the flux through the Sun's photosphere (64 MW/m^2) or through a rocket nozzle (7,000 W$_t$/cm^2). Standard operating densities on the order of 100 W$_t$/cm^2 are of the same order of magnitude as the heat rejection of hot flue gases through a large stack, and far higher than the heat generated by the US Space Shuttle's reentry into the atmosphere, about 6 W$_t$/cm^2 for thermal protection tiles and about 60 W$_t$/cm^2 for the leading edge (Harvey 2008). Subsequent microprocessor redesigns reduced the heat flux, in 2006 the Core 2 Duo (65 W) to below 50 W$_t$/cm^2, but by 2010 Intel's Atom was back to the hot-plate power density of 100 W$_t$/m^2 (Pant 2011), and the power density at 1-mm^2 hot spots can be higher than 300 W/cm^2.

Many ingenious arrangements are needed to manage high heat fluxes (Allan 2011). This challenge becomes even greater when servers and disk storage systems are stacked in data centers in racks. A standard server rack, placed on a raised floor with perforated tiles or grates, has been 2.1 m tall, composed of forty-two units of 48.26-cm (19-inch)-wide slots, but some companies have preferred much taller racks (fifty-seven units by Microsoft) or wider (up to 58.42 cm) enclosures (Miller 2012). The footprint of a

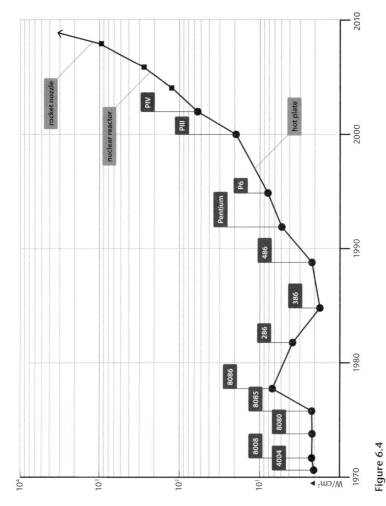

Figure 6.4

Heat rejection by microprocessors. Carl De Torres Graphic Design. Based on data in Azar 2000, Joshi 2001, and Intel 2005.

standard 42-unit rack is 0.478 m^2 for the inside dimensions and 0.65 m^2 for the cabinet. The rate of increase in product heat density has slowed since the mid-2000s, but between 1990 and 2010 servers and disk storage systems saw an order of magnitude rise, from just over 2,000 W/m^2 to 20,000 W/m^2 (Data Clean 2012). Most of the products deployed in large data centers have a rated demand between 8 and 20 kW per rack, and designers are working on servers of 30 kW+.

Other space definitions reduce this density: adding front and back clearances around server cabinets cuts the density by roughly two-thirds, adding shared aisles at the end of the row entails a further 25% cut, and including all main aisles and power delivery and cooling units brings the rate down to less than 20% of the rack density value (Brill 2006). The approximate conversion of product (rack) power density to gross computer room density is just 0.018. Moreover, some racks may be empty (held in reserve), while others may be only partially filled, and the centers rarely operate near their full capacity. On the other hand, for every 1 W used by servers up to 0.9 W have typically been needed for the electrical infrastructure that enables the servers, that is, for an uninterruptible power supply, switches, lighting, and, above all (nearly 40% of all power), for cooling (Emerson Network Power 2009).

We have representative data on power densities for data centers housing the servers based on information from 59 facilities in North America and Europe with areas ranging from less than 2,000 m^2 to 18,000 m^2 (Renaud and Mescall 2011). The centers were designed for a median power density of 730 W/m^2, with some smaller centers rated for as much as 1,615 W/m^2, but actually used power (gross computer room density) averaged only 437 W/m^2, which means that the centers are using less than 60% of their available power. Their median value per rack was only 2,100 W/m^2, with the middle 50% ranging from 1,700 to 2,200 W/m^2, and the average power density per rack declined with the data center size. Some smaller centers may have overall densities on the order of 2,500 W/m^2 (Emerson Network Power 2008).

Coping with less than 500 W/m^2 in terms of gross computer room density is not an extraordinary challenge, but removing heat from racks operating with densities of 10^3 W/m^2 (and for rooms whose power density is in excess of 2,000 W/m^2) calls for efficient cooling to maintain the temperature between 10°C and 35°C and the relative humidity below 80% (Minas and Ellison 2009; Pflueger 2010). This challenge is perhaps best expressed by the rack power densities in terms of power/volume. A high-rise apartment building in Hong Kong may have a power density twice as high as a server room when both rates are expressed per unit of their footprint area (950 vs. 475 W/m^2), but a server rack using 2,100 W/m^2 will be rejecting about 1,000 W/m^3 of its volume, compared to less than 10 W/m^3 of a 38-story apartment building.

Uneven cooling, and hence recurrent hot spots, are a common problem in data centers, with about 10% of racks getting too hot to meet the standards for maximum reliability: every increase of 10°C above 20°C cuts the long-term reliability of hardware by half (Renaud and Mescall 2011). The cooling requirements of data centers have been rising, but 70% of cooling capacity available in a typical computer room is wasted due to bypass airflow. The global electricity demand by data centers doubled between 2000 and 2005 (Koomey 2008), but between 2005 and 2010 (due to recession) it increased by only about 56%, and in the United States it grew by just 36% (Koomey 2011). In aggregate terms, data centers used 1.1%–1.5% of the world's electricity (23–31 GW) in 2010, and the US share was between 1.7% and 2% (7.7–9.8 GW). But between 2011 and 2012 global power requirements rose by 63%, from 24 GW to 38 GW, with 43 GW forecast for 2013 (*Computer Weekly* 2012). Given a rapid data growth—Intel sees new information load rising from 2 zettabytes (10^{21} bytes) in 2011 to 8 petabytes in 2015 (Otellini 2012)—further large increases in demand must be expected.

Finally, a few observations on the thermal consequences of energy use in general, and on concentrated heat rejection in particular, are in order. Proceeding from the smallest to the largest scale, it is obvious that the highest power densities of heat rejection are limited to such tiny areas—as small as 10^{-4} m^2 for microprocessors—that the power fluxes of individual components are too small to have any discernible impact even on a room scale. Keeping a microchip cool may be a challenge, and careful design must be deployed to cool microprocessors in servers that are layered in tall racks— but it is a challenge only in rooms crammed with data processing assemblies where their operation dominates the overall heat balance.

In residential settings, heat rejected by electronic devices represents a negligible addition to a room's thermal background. When I am using a laptop in my living room its operation will add some 60 W of dissipated heat, compared to 400 W coming from recessed ceiling lights, 200 W from a stereo system, 85 W from my metabolism (assuming 1.2 times the basal metabolic rate while sitting)—and, during a summer day, more than 1,500 W of sunlight that is coming in at noon through southwest-facing windows, or, during cold winter evenings, about 1,000 W of hot air that is forced through floor heating vents.

Anthropogenic heat rejection rates that are three orders of magnitude above the peak insolation rates (that is, 10^6 W/m^2) are restricted to areas smaller than 10^2 m^2 (the bottoms of large power plant boilers, the mouths of tall power plants stacks), and those flows that are one to two orders of magnitude higher than noontime irradiance (that is, between10^4 and 10^5 W/m^2) originate from areas smaller than 100 by 100 m, mainly large cooling towers. Heat rejection by both tall power plant stacks and cooling towers is associated with substantial condensation, often with the formation of clouds, and with some recurrent local fogging and icing, but only rarely with any downwind precipitation anomalies.

Other concentrated heat rejection processes (ranging from hot plates and car exhausts to tall power plant stacks) are limited to relatively small areas: as the power density of anthropogenic energy conversions increases, the spatial extent of the more intensive heat rejection fluxes declines at a considerably faster rate. As a result, heat-rejection phenomena that constitute a significant share of solar inputs (10^1 W/m^2) are limited to areas no larger than 10^8 m^2, to large cities, extensive industrial regions, and busy transportation corridors. As already noted, energy use in buildings, industries, and transport helps turn large cities into urban heat islands. Other factors contribute to this phenomenon: impervious surfaces have a higher thermal capacity than does vegetation, and their heating generates stronger convective flows; many buildings and most parking lots and roads have a lower albedo than many natural surfaces; and the restricted sky view in canyon-like streets reduces radiative cooling and sensible heat loss is larger than latent heat flux (Stewart 2011).

As a result, urban heat islands commonly average 2°C more than the surrounding countryside, and peak differences may be temporarily (especially during the night) as much as 8°C higher. Urban heat islands can explain only a negligible fraction of global temperature rise during the twentieth century (Peterson and Owen 2005), but they have a number of well-documented impacts, including statistically significant (local and downwind) increases in cloudiness, precipitation, and thunderstorms and reductions in relative humidity, wind speed, and horizontal insolation as a result of shading. They also promote the formation of photochemical smog, contribute to premature mortality during summer heat waves (Wong, Paddon, and Jimenez 2013), and increase energy use for air conditioning.

7   Making Sense of Power Densities

A brief recapitulation of power densities representing all major energy conversions and uses precedes some revealing comparisons of supply and demand. The chapter closes with summations of land claims by the global energy systems and by the US provision of fossil, nuclear, and renewable energies.

My recapitulation will avoid both a perfunctory recital of a few basic generalizations and an elaborate account of all noteworthy conclusions. Instead, I will offer a systematic but concise review of key findings, and then present comparisons of the typical power densities of specific production activities and major energy uses. I will also use representative power density means to calculate approximate aggregates of the spatial claims of energy systems, first on a global level (to set energy provision in a wider context of changing land use), then, with a greater degree of accuracy, for the United States. But before I do so, some analytical points are in order.

Higher power densities mean lower land claims, but the highest possible densities are not always most desirable. Well-designed surface coal mines, with surfaces reshaped and land returned to agricultural, forestry, or recreational use in less than a generation, are preferable to underground mining with its occupational hazards and often prolonged periods of ground subsidence. And while run-of-river plants would be the most desirable choice (as they do not claim any additional land), they have obvious disadvantages: unlike PV-based solar and wind generation of electricity, they may not cease to operate altogether, but their capacities may be periodically limited in all regions with widely fluctuating water flows; in contrast, space-intensive large reservoirs ensure a predictable capacity far above that corresponding to the minimum stream flow.

And, most fundamentally, qualitative differences invalidate any simplistic verdicts about superiority. High-power-density fossil energies—which are also cheaper to transport, easier to store, and can be used on demand—are preferable but not inherently superior because, unlike solar radiation and its conversions, they are exhaustible on the civilizational time scale of 10^3 years. And while many conversions of fossil fuels are now virtually pollution-free, removing and storing CO_2 at scales that would eliminate the gas as a key factor of anthropogenic climate change (a necessity largely eliminated with renewable sources of energy) remains an extraordinary challenge. At the same time, thermal electricity generation has unrivaled availability, a key consideration in societies dependent on an incessant and highly reliable (99.9999% of all time) supply. The most commonly encountered capacity factors range from 10% to 15% for PV-based electricity generation in mid-latitude locations and 15%–25% in the sunniest environments to 20%–35% for wind farms in moderately to very windy regions, 40%–60% for large hydro stations, 60%–85% for coal-fired electricity-generating plants, and about 90% for nuclear reactors.

Recapitulations

Modern civilization energizes its complex metabolism by converting two basic kinds of resources into useful energies: renewable flows, and finite stores of fossil fuels and uranium. Except for geothermal and tidal energies, all renewable flows are transformations of a tiny fragment of solar radiation that reaches the biosphere. Nearly a third of that incoming flux is reflected by the Earth's clouds and surfaces, and nearly all the rest is reradiated, with only a small share driving atmospheric circulation (of which a small share can be harnessed as wind energy), the global water cycle (with river flows partially convertible to hydro-generated electricity), and photosynthesis, the source of biofuels and fossil fuels (after 10^5–10^8 years of heat and pressure processing).

Renewable Flows
Solar radiation flux is so large that even fairly low conversion efficiencies—to heat (mostly in rooftop water heaters) or to electricity (by PV cells or by central solar plants)—can harness it with relatively high power densities. For heat, the maxima are on the order of 10^2 W/m^2; for PV-based electricity

generation, they reach 10^1 W/m^2. Power densities for most hydro stations are an order of magnitude lower (10^0 W/m^2), large-scale harnessing of wind does not go above 10^0 W/m^2, exceptionally high phytomass harvests yield power densities a bit above 1 W/m^2, and harvests of wood and crops can be converted to useful energies with power densities of just 10^{-1} W/m^2.

During noontime hours the power densities of solar water heating in sunny climates can surpass 700 W/m^2, by far the highest rate of any renewable energy conversion. Annual power densities are on the order of 100 W$_t$/m^2 in sunny subtropical climates and 40–50 W$_t$/m^2 in cloudy mid-latitudes, while the global mean (based on actual performance) was 67 W$_t$/m^2 in 2012. Two common biases must be avoided in order to calculate comparable power densities of PV-based electricity generation: the rates should not be expressed in terms of peak power and the area should not include only module surfaces. The power densities of PV-based electricity generation decline by an order of magnitude as we proceed from noontime maxima to annual averages and from panel areas to total areas of PV projects. Noontime maxima at prevailing efficiencies (10%–15%) are 80–150 W/m^2 of a module, and modules cover 75%–80% of an actual PV field and 25%–75% of a total solar park area. Capacity factors correlate with irradiance and range from less than 12% to 25%.

Even with today's still fairly low conversion efficiencies, solar PV has by far the highest power density of all practical options for electricity generation based on new renewable flows. As further cost reductions and further efficiency gains are certain, the development of this source should receive commensurate attention. Large ground-mounted PV projects now generate electricity with power densities of between 3 and 7 W$_e$/m^2 in less sunny locations and 7–11 W$_e$/m^2 in sunny regions (all rates are for the total plant area). Germany has the largest area of rooftop modules, and the power density of their PV-based electricity generation averages 12 W/m^2 of roof area covered by solar panels. The power density of PV panel–covered walls will be always lower, in most instances less than 5 W/m^2.

Continuing efficiency gains should soon raise power densities above 10 W/m^2 common in large-scale PV projects in sunny locations, and it is not unreasonable to expect rates well above 20 W/m^2 by mid-century; should they turn out be commercially viable, three-dimensional PV converters could push the values far above 50 W/m^2. Pioneering central solar power projects have power densities (calculated for their total area) between 4 and

6 W/m^2, similar to those of PV conversions. The largest new design in the California desert averages almost 50 W/m^2 for the heliostat area and nearly 9 W/m^2 for the entire project.

Most of the planet's large wind energy potential is at high altitudes, where engineering challenges and economic realities preclude any imminent commercial conversions. Electricity generation by ground-based wind turbines has typical spacing power densities on the order of 10^0 W/m^2 in terms of installed capacity and 10^{-1} W/m^2 for actual annual production. Large (3–4 MW) turbines set in a square grid have power densities of mostly 2–3 W$_i$/m^2, and the actual range for America's large wind farms is roughly 1–11 W$_i$/m^2, with most projects between 2.5 and 4 W$_i$/m^2. The annual capacity factors of wind generation have been rising, with recent nationwide means between 15% (Germany) and 25% (UK) in Europe, and the US maximum reached 33%. Average operating power densities are thus less than 0.75 W/m^2 in Europe and about 1 W/m^2 in the United States, an order of magnitude below that of PV-based electricity generation. When only the tower footprints and access roads are counted, the power densities of wind generation rise by an order of magnitude, but that is an erroneous value to use in making comparisons with other modes of electricity generation because it does not convey the fundamental spacing requirements that limit the power densities of wind turbines.

Although power densities of large hydro projects span four orders of magnitude, from 10^{-1} to 10^2 W/m^2, most of the world's hydroelectricity comes from plants whose installed capacities and large reservoirs translate to rates of 10^1 W$_i$/m^2. The average power densities of hydro projects with capacities of less than 100 MW are below 0.5 W$_i$/m^2; the mean is nearly 1.5 W$_i$/m^2 for stations up to 1 GW, and it surpasses 3 W$_i$/m^2 for projects in excess of 3 GW. This correlation of rising densities with rising capacities continues, as the projects with the world's largest installed capacities (China's Three Gorges Dam, Brazil's Itaipu, the Grand Coulee Dam in the United States, all above 6 GW) have power densities between 10 and 20 W$_i$/m^2. As expected, the highest power densities, some in excess of 500 W$_i$/m^2, belong to some alpine stations with high heads and small reservoirs.

The power densities of actual electricity generation are considerably lower because the typical capacity factors of hydro stations are usually less than 60%, and often less than 50%—and even less when used just for

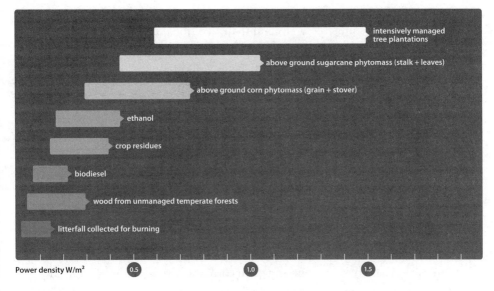

Figure 7.1
Power densities of phytomass and biofuel production. Carl De Torres Graphic
Design.

peaking power. Actual operating power densities are thus roughly halved,
with most of the rates falling to less than 10 W_e/m^2 for the largest stations,
less than 3 W_e/m^2 for medium-sized projects, and less than 1 W_e/m^2 for stations with the largest reservoirs and low capacity factors. Hydroelectric generation is thus highly space-demanding, with many smaller projects having
power densities of just around 1 W_e/m^2 or less, similar to those of phytomass harvests in temperate climates, but with the global mean pushed up
by higher power densities (above 5 W_e/m^2) for the largest stations that supply most of the world's hydroelectricity.

Photosynthesis has an inherently low efficiency, and the power densities
of phytomass harvests are highly correlated with temperature and precipitation during the growing season and with the availability of macro- and
micronutrients (fig. 7.1). Dense plantings of fast-growing trees harvested in
short rotations have very high yields in small experimental plots but will
not perform that well in large plantations with no or only moderate fertilization and without irrigation. Realistic yields in most temperate environments are 5–15 t/ha (power densities of 0.3–0.9 W/m^2) and 20–25 t/ha in
the subtropics and tropics (power densities of 1.2–1.5 W/m^2).

The modern commercial version of charcoal making (in Brazil, to supply blast furnaces) is much less wasteful than traditional methods were but its power density is still no higher than about 0.6 W/m^2; higher tree plantation yields and increased charcoaling efficiency could raise it close to 1 W/m^2. Converting a rich harvest of tropical trees (20 t/ha or 1.2 W/m^2) by three different methods—by the combustion of woody phytomass (an efficiency of roughly 90%), by wood gasification and subsequent combustion in engines or turbines (an overall efficiency of 35%–40%), and by the production of methanol (an efficiency of 70%)—would result in power densities of, respectively, 1.1 W/m^2, about 1 W/m^2, and around 0.8 W/m^2, and the density would be less than 0.5 W/m^2 if wood-based gas were to be used for electricity generation. For wood harvests from temperate forests (10 t/ha, 0.6 W/m^2), all of these values would be roughly halved.

The production of liquid biofuels from crops is even more space-intensive. Yields of sugar cane and grain corn, the two most important feedstock crops, have been rising (as have the efficiencies of their fermentation), but the power density of producing ethanol from US corn yielding 10 t/ha is no higher than 0.25 W/m^2. In Asia, where yields are much lower, the power density of corn ethanol production would be just above 0.1 W/m^2. In contrast, the power density of ethanol made from high-yielding Brazilian sugar cane is about 0.4 W/m^2, and cellulosic ethanol, as yet to be commercialized on a large scale, would have a similar power density.

The production of biodiesel from oil seeds is limited by the inherently low yields of oil crops. Even for the relatively high-yielding Dutch rapeseed, the power density is only 0.18 W/m^2, and the EU's biodiesel mean is only 0.12 W/m^2. Finally, the small-scale conversion of biomass wastes to produce biogas is rather inefficient: even the best commercial conversion, using corn silage as feedstock, translates to power densities of about 0.6 W/m^2 for the gas and 0.2 W/m^2 for using that gas to generate electricity. The inherent limits of a phytomass-based energy supply mean that even an unlikely early doubling of yields would keep the power densities of wood-based electricity generation and of ethanol, biodiesel, and biogas production mostly below 1 W/m^2.

Large geothermal projects have power densities between 100 and 800 W$_i$/m^2 of directly affected land. These values go down an order of magnitude, mostly to 50–80 W$_e$/m^2, when actual generation is prorated over all affected land of accurately assessed projects in the United States, Iceland,

and New Zealand. The power densities of geothermal electricity generation are thus similar to those of alpine stations, an order of magnitude above typical PV performance in Atlantic Europe (5 W_e/m^2) or in the more sunny United States (8–10 W_e/m^2), and also an order of magnitude higher than the spacing densities for the best large wind farms. Supplies of geothermal heat for individual houses have similar densities, commonly between 40 and 100 W_t/m^2.

Fossil Fuels

Compared to the sprawling land claims of phytomass energies or hydro-electric power generation, extraction of the richest deposits of fossil fuels is almost punctiform; at the same time, the power densities of fossil fuel extraction show even wider ranges than do the conversions of renewable energies. The specific rates of coal mining depend on the depth, thickness, and quality of coal seams. The permanent structures of large underground mines (machinery, maintenance, storage, and office buildings, parking lots, coal processing and shipping facilities) occupy relatively small areas (on the order of 1 ha/1 Mt of annual output); much more land is taken by the on-site disposal of rocks separated from coal and by tailing ponds that receive small-particle waste. The largest underground mines producing high-quality coal from thick seams have power densities in excess of 10,000 W/m^2, while the rates for smaller operations extracting lower-quality fuel from thinner seams (and hence generating more waste) are often above 2,000 W/m^2 and usually not below 1,000 W/m^2.

The ratio between variable and fixed land claims is much larger for surface mining because large volumes of overburden must be stripped and repositioned to access coal seams. The highest ratios of overburden to coal seam are now nearly 7, and the deepest surface mines go below 300 m. Many destructive and unsightly mountaintop removals in central Appalachia produce coal with power densities of just 200 W/m^2, some well below 100 W/m^2, while the operating densities of the largest surface mines that extract the thickest seams (10^2 m), be it in Wyoming's Powder River Basin or Australia's Latrobe Valley, are more than 10,000 W/m^2 and even above 15,000 W/m^2.

The lowest power densities of coal extraction are thus comparable to those of PV-based electricity generation, while the highest rates are as high as hydrocarbon production in some major oil and gas fields. Most surface mines do not belong to either of these extreme categories: their operating densities are typically 1,000–5,000 W/m^2, and the figures do not change significantly even when dedicated railroad links (built solely to move coal from a large mine to a power plant) are included because land disturbed by surface mining over the lifetime of an operation greatly surpasses the rights-of-way of a typical rail line.

Early oil field development was often marked by drilling too many wells in close proximity in a predatory quest for maximized output; modern development optimizes production, with 5–30 wells/km^2 and with as few as 2–3/km^2 where multiple directional or horizontal wells can be drilled from a single pad. A small number of giant oil reservoirs produce a disproportionate amount of oil, but their lifelong power densities cannot be accurately calculated: the largest reservoirs are still producing, with no firm numbers on their ultimate aggregate output. North American oil fields show long-term cumulative power densities of about 2,500 W/m^2 for more than 80 years in California and about 1,100 W/m^2 for 50 years in Alberta, with both rates calculated for land occupied by wells.

Annual statistics make it possible to trace a gradual decline in power densities (assuming, liberally, 2 ha/well). Between 1972 and 2012 they dropped from about 40,000 W/m^2 to 23,000 W/m^2 in Saudi Arabia, and from almost 25,000 W/m^2 to less than 9,000 W/m^2 in the Middle East. Very low US and global rates (100 and 650 W/m^2) are biased by the inclusion of thousands of America's old marginal wells, kept in production because of recent high oil prices. The span is thus between 10^2 W/m^2 for mature fields that have been producing for generations to 10^4 W/m^2 for the world's most productive reservoirs, with modal values near the lower end of 10^3 W/m^2.

The two most important ways of nonconventional oil production are horizontal drilling and hydraulic fracturing of oil-bearing shales in the United States, and extraction of oil from Alberta's sands. Oil shale wells have rapid, hyperbolic productivity declines. For North Dakota's Bakken shale, now the largest oil shale producing region, power densities decline from about 4,000 W/m^2 in the first year to a mean of 1,600 W/m^2 for the first five years and less than 1,000 W/m^2 for ten years of extraction. The production of oil from Alberta's oil sands began with surface mining of sands; the power densities of these operations are similar to those of surface coal mining, ranging from roughly 2,000 to 4,000 W/m^2. In contrast, in situ recovery, the dominant way of future production, has power densities of 7,000–16,000 W/m^2 and averages about 10,000 W/m^2. These high densities drop significantly with inclusion of the land required to produce natural gas use for bitumen extraction and steam generation: the power densities of surface mining fall to about 2,300 W/m^2 and those of in situ extraction to less than 3,200 W/m^2.

The transport and processing of crude oil have fairly high throughput power densities. Crude oil pipelines need construction corridors 15–30 m wide, and ROWs typically are 15 m for buried lines. The average US throughput power density is nearly 700 W/m^2, with major trunk lines rating well above 1,000 W/m^2. Pumping of crude oil into large tankers has very high throughput power densities of 10^5 W/m^2. A disproportionate amount of liquid fuels comes from a relatively small number of large refineries whose processing power densities are mostly between 4,000 and 8,000 W/m^2.

The power densities of natural gas and crude oil extraction are similar, with most operations rating between 10^3 and 10^4 W/m^2: 2,300 W/m^2 for all of Alberta's extraction to nearly 50,000 W/m^2 for Groningen, the Dutch supergiant field. The gas output from hydraulically fractured horizontal wells shows the same rapid decline as oil production from shales, from power densities of 10^3 W/m^2 in the first year to a low of 10^2 W/m^2 just a few years later. The processing of natural gas has high throughput densities of 10^4 W/m^2, and the power densities of long-distance pipeline transportation range from 10^2 to 10^3 W/m^2. The rising trade in liquefied natural gas (LNG) involves high-power-density operations: gas liquefaction proceeds at a high power density of 10^3 W/m^2, while the power densities of regasification can be well into 10^4 W/m^2.

Thermal Electricity Generation

The steam-driven generation of electricity remains the dominant way of producing the most flexible kind of energy, and the two most important ways to produce steam are by burning pulverized coal in boilers and by fissioning uranium in nuclear reactors. Burning gas in boilers is much less common because gas turbines are much more efficient (as well as more flexible), and using crude oil or refined liquid fuels for electricity generations is rarer still (because liquids are too expensive). The core structures of thermal stations are quite compact, and plant sites are mostly occupied by essential infrastructures. Many plants also hold additional land in reserve for possible expansion or as a buffer zone, and nuclear plants require a safety belt that excludes permanent habitation. Some thermal electricity generation proceeds with a very high power density, but plants with large adjacent cooling lakes, with large areas reserved for on-site storage of captured fly ash and sulfates from flue-gas desulfurization, and with extensive (often forested) buffer zones have much lower ratings.

Thermal electricity generation is clearly in a high-power-density category, but its surprisingly large range of values means that its outcomes can be orders of magnitude above any renewable alternatives (mine-mouth coal-fired stations burning high-quality fuel, gas-fired plants burning LNG)—or that it can have almost as low a power density as the best instances of large-scale PV-based generation (coal-fired stations burning low-quality fuel from surface mines with a high overburden to seam ratio). Here is the sequence of diminishing power densities for coal-fired electricity generation, from its burning core to complete plant-and-fuel claims.

Large boilers have power densities in the low 10^6 W_t/m^2 of their footprint; boilers and turbogenerator halls are in the mid- to high 10^5 W_e/m^2 range; stations, including all on-site infrastructures, are in the mid-10^3 W_e/m^2 range; and stations including coal extraction and delivery infrastructure are in the low 10^3 W_e/m^2 or high 10^2 W_e/m^2 range (determined by coal quality and shipping distance). Actual generated power densities for coal-fired stations in the United States, Europe, and Asia confirm model calculations, as they range from a high 10^2 W_e/m^2 to a mostly low 10^3 W_e/m^2 range for plant sites only; depending on the mining methods, coal quality, and coal shipping; inclusion of the entire coal-to-electricity chain leaves the densities well above 1,000 W_e/m^2 or lowers them to the mid-10^2 W_e/m^2 range, even to just above 100 W_e/m^2.

Stations burning heavy oil and crude oil have operating power densities typically well into the 10^3 W_e/m^2 range. Natural gas–fired central electricity-generating plants do not need any land for extensive on-site waste deposits, and continuing gas supply by pipelines or LNG tankers reduces their fuel storage requirements; thus they generate electricity with power densities mostly between 2,000 and 6,000 W_e/m^2. Even after the claims of natural gas production are included, the power densities for the entire extraction-generation sequence remain on the order of 10^3 W_e/m^2. More natural gas is now burned in gas turbines than in the boilers of large central stations. Gas

•

turbines have limited ratings (10^0 to mid-10^2 MW) but operate with very high power densities, commonly 10^4 W_e/m^2. This makes it possible to site new gas turbines on land belonging to the existing central stations. The same is obviously possible for combined-cycle plants, whose operating power densities will also be on the order of 10^3 W_e/m^2.

The power densities of nuclear reactors (per unit of their footprint) are 10^6 W_t/m^2, but subsequent energy flows are the same as in fossil-fueled central stations, and hence the layout of nuclear plants and their overall land requirements are also very similar. Because these plants do not require any extensive fuel storage and do not generate any waste from controlling air pollutants, they can be quite compact, rating between 2,000 and 6,000 W_e/m^2. Land claims for the entire uranium cycle lower the densities of the entire fuel generation-disposal sequence. The most productive underground mines have exceptionally high extraction densities (10^4 W_e/m^2), typical surface operations produce ore with power densities ranging from the high 10^2 W_e/m^2 range to the low 10^3 W_e/m^2 range, and in situ leaching projects reach 400–600 W_e/m^2.

The milling of ores to produce U_3O_8 concentrate proceeds with high power densities, and the production of fuel rods filled with slightly enriched UO_2 is even less space-intensive, but the enrichment process has fairly low power densities once the claims for its electricity requirements are included. Differences in uranium mining, processing, fuel enrichment, and eventual radioactive waste disposal produce a range of overall power densities very similar to that of coal-fired plants, with rates as high as 10^3 W_e/m^2 and as low as the low to mid-10^2 W_e/m^2 range. Several recent studies offer all-inclusive rates of 230 to 960 W_e/m^2 for US nuclear power generation, in good agreement with my detailed, stepwise quantification.

Early electricity-generating plants served their immediate neighborhoods, but modern stations are connected to load centers (cities, industries) by high-voltage AC transmission lines. Their ROWs add up to mostly between 4 and 6 ha/km, while high-voltage DC links (mostly from distant hydro stations) need at least 40% less land for handling the same capacity. Naturally, short, high-capacity links will have higher power densities—a 200-km, 1-GW, 765-kV line would rate about 80 W/m^2 of its ROW—but the means for large utilities or for nationwide grids will typically be no more than 30–40 W/m^2 of ROW.

Energy Uses

As huge as it is in absolute terms (17 TW in 2013), global energy use remains a tiny fraction of the Earth's natural radiation balance: it prorates to 0.03 W/m^2 of the Earth's surface and to 0.125 W/m^2 W/m^2 of ice-free land, the latter rate being less than 0.07% of the mean global insolation and a small flux compared to the 2.3 W/m^2 of aggregate radiative forcing due to greenhouse gases. Average power densities of energy use within national territories range over four orders of magnitude: desert countries with sparse rural population rate only 10^{-4} W/m^2, while the Netherlands, Belgium, and South Korea reach about 3 W/m^2 (fig. 7.2). Densely populated and industrialized urban areas use about 75% of the world's energy, and the global average of their power density reaches roughly 20 W/m^2.

That, too, is roughly the US urban mean, including all treed and grassed area within cities; after their exclusion the power density per unit of impervious US urban surfaces is about 35 W/m^2. Worldwide, annual urban means range between 10 and 100 W/m^2; downtowns average in excess of 100 W/m^2 and their hourly extremes often approach and can substantially surpass 1,000 W/m^2, and that much can be the annual mean for the densest city blocks in Manhattan, a flux equaling or exceeding noontime insolation. The variability of rates among individual buildings in the same climate is highly influenced by their function and construction.

The power densities of detached houses and apartments now differ less than just a generation or two ago because the ownership of many household energy converters (refrigerators, electric or gas stoves, washing machines, TVs, computers) has approached or reached saturation. Moreover, higher efficiencies have cut the household rate in North America and Europe by half or more since the 1960s. The most efficient commercial buildings need less than 10 W/m^2 of floor area, a tenth of the 1970s level. But in temperate climates, with winter heating and summer cooling, power densities are around 20–30 W/m^2 of floor area in single- and multifamily housing. Schools have similar rates. Parking garages and warehouses have the lowest power densities, hospitals the highest (about 70 W/m^2 of floor area). Adjustments from floor area to building footprints yield some very high power densities, up to roughly 1,000 W/m^2 for the nearly 40-story housing estate towers in Hong Kong, 2,000 W/m^2 for a 50-story luxury hotel in a hot climate, and more than 6,000 W/m^2 for Burj Khalifa, the world's tallest building.

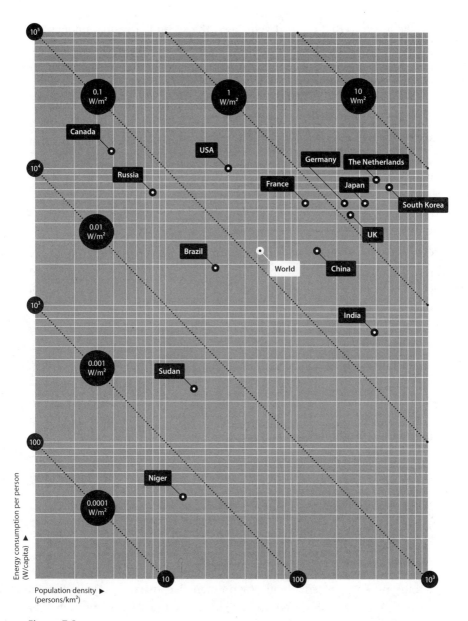

Figure 7.2
Power densities of large-scale energy use. Carl De Torres Graphic Design. *Notes:* Population density (persons/km^2) is displayed on the *x*-axis, energy consumption (W/capita) on the *y*-axis. Power density (W/m^2) is arrayed on the diagonals.

Energy converted by road vehicles is the second largest component of urban fuel use. The free-flowing traffic of cars and trucks will have a short-term power density of about 500 W/m^2 of a paved lane, and this figure may nearly double when vehicles have to idle. When the vehicular energy use is prorated over the entire ROW (lanes, shoulders, and medians), the densities are roughly halved (and cut even more in rural settings), and a further major reduction comes when parking lot space is included. For entire urban areas, with traffic averaged on an annual basis, the rates are usually less than 5 W/m^2.

Comparisons

Comparisons of power densities inform us in two important ways: they reveal the hierarchies of supply and use, and the contrasts of supply and usage rates allow us to quantify relative claims of particular energy requirements and to assess the future demands for land devoted to energy industries. Of course, we should keep in mind that only a few power densities have very narrow ranges, and that in many specific cases the ratings will be far from the quoted typical performances or well outside the usual ranges. Perhaps the most consequential reality is that the extraction and conversion of fossil fuels and uranium produce useful heat and electricity with power densities that are usually at least two and up to five orders of magnitude higher than the exploitation of renewable flows. Only in exceptional cases do some renewable energy conversions (most notably those involving alpine hydro stations with high heads and small reservoirs) have higher production power densities than the extraction of fossil fuels and the generation of thermal electricity (fig. 7.3).

The production of phytomass in general, and of liquid biofuels in particular, has the lowest power densities of all commercially exploited resources. This is a fundamentally unalterable outcome of inherently low photosynthetic efficiency and environmental constraints on phytomass yield. Crops can be produced with a higher power density than woody biomass: the latter has a higher energy density, but the former has a higher rate of growth. Although yields have increased for both field crops and plantation trees, and although further increases are coming, there is no realistic prospect for biofuels to be produced with a power density significantly surpassing 0.5 W/m^2 in temperate environments and 1 W/m^2 in tropical

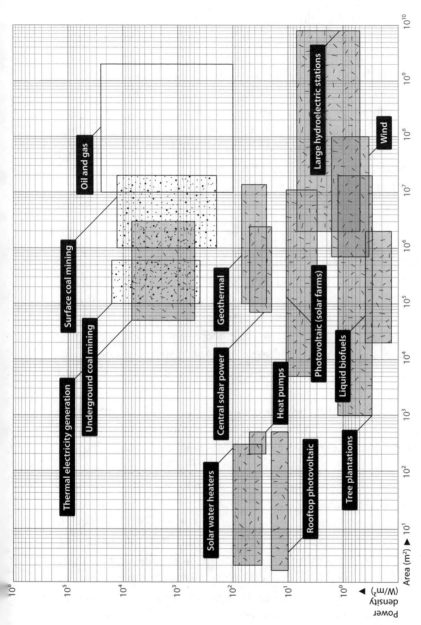

Figure 7.3

Power densities of fossil energy conversions and renewable energies. Carl De Torres Graphic Design. *Notes:* Area (m²) is displayed on the *x*-axis, power density (W/m²) on the *y*-axis. Figures were calculated for liquid biofuels, tree plantations, wind energy, large hydroelectric stations, PV-based sources (solar farms), rooftop PV arrays, central solar power, geothermal power, heat pumps, solar water heaters, oil and gas operations, surface coal mining, thermal electricity generation, and underground coal mining.

environments, where the current rates are less than half these thresholds. Moreover, if the inputs needed to produce biofuels (machinery, fuel, fertilizer) were to be energized solely by renewable energies rather than (as is the case) by fossil fuels and nuclear-derived electricity, then the power densities of such completely renewable operations would drop to less than 0.1 W/m^2.

Hydroelectric power projects with large dams and low capacity factors can be as land-intensive as phytomass production, while some of the world's largest plants and alpine-type stations can generate electricity with power densities up to two orders of magnitude higher. The high land requirements of hydroelectric power generation are best illustrated by the fact that this conversion supplies less than 3% of the world's primary energy but accounts for more than half of all land occupied by the world's energy infrastructure. Wind turbine spacing results in power densities hardly better than the best phytomass production, but when only actual (physical) land claims are counted the rate goes up by an order of magnitude and, depending on location, rivals or surpasses the power densities of PV-based solar generation. Central solar power is not significantly less land-intensive but has much higher capacity factors. No other mode among today's commercial conversions of renewable energy can do better than flat-plate solar collectors, which can supply hot water with power densities between 50 and 100 W$_t$/m^2.

Extreme values for the power densities of coal extraction span three orders of magnitude, from less than 100 W/m^2 to more than 1,000 W/m^2, but typical rates in large modern surface mines are very similar to those in large oil fields, being at least close to 1,000 W/m^2 and commonly two to four times higher. Coal and natural gas extraction make up the largest land claims of fossil fuel–based thermal electricity generation, and while the plants themselves are fairly compact, a great deal of space may be claimed by cooling and captured waste disposal infrastructure. This may reduce the power densities of some coal-fired stations to the same order of magnitude as those of large solar PV-based electricity generation, while the fastest-growing choice, gas-powered turbines, is highly compact, with power densities of 10^3 W$_e$/m^2. The power densities of nuclear generation are broadly comparable to those of coal-fired power plants (10^2–10^3 W$_e$/m^2).

The power densities of energy production range over five orders of magnitude, from 10^{-1} W/m^2 for liquid biofuels to 10^4 W/m^2 for the world's richest hydrocarbon deposits, but the final energy uses of modern high-energy societies fall mostly between 10^1 and 10^2 W/m^2 for homes, commercial buildings, industrial enterprises, and densely populated urban areas. *This means that modern civilization extracts fuels and generates thermal electricity with power densities that are commonly at least one, usually two, and sometimes three orders of magnitude higher than the power densities of final energy uses in urban areas (where most people now live) and in individual buildings and commercial and industrial establishments* (fig. 7.4).

Fossil fuels to supply urban areas are extracted and delivered with power densities that are higher than the power densities of large cities (10–30 W/m^2). Thermal electricity is typically generated with power densities that are one and often two orders of magnitude higher (300–3,000 W$_e$/m^2) than the power densities of electricity use in family houses (10–50 W$_e$/m^2). Liquid fuels for transportation are produced with power densities that are one to two orders of magnitude higher than the power densities of urban traffic. And even the very high power densities (300–1,000 W/m^2 for supermarkets, high-rises, factories, and downtowns) either overlap or are slightly surpassed by the power densities with which electricity and fuels are actually produced and delivered.

The modern energy system produces concentrated energy flows and then diffuses them through pipelines, railways, and high-voltage transmission lines to final users. As a result, the space claimed by the extraction and conversion of fossil fuels is a small fraction of the ROWs needed to distribute fuels and electricity: American extraction, processing, and conversion of coals and hydrocarbons take up less than 20% of the land that is required for pipeline, railway, and transmission ROWs and occupy less than 0.1% of the country's territory. *In contrast, future societies powered solely or largely by renewable energies would rely on an opposite approach by concentrating diffuse energy flows captured with low power densities ranging mostly between 0.2 W/m^2 for liquid biofuels to 20 W/m^2 for solar PV-based energy. Renewable energy systems would have to bridge gaps of several orders of magnitude between the power densities of energy production and use* (fig. 7.5).

As a result, tomorrow's societies, which will inherit today's housing, commercial, industrial, and transportation infrastructures, will need at least

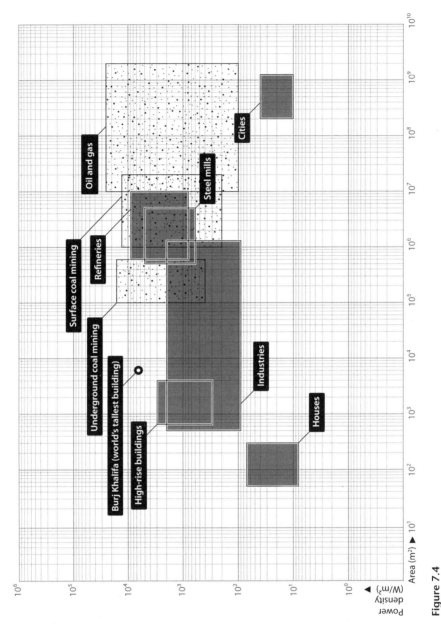

Figure 7.4
Power densities of fossil fuel energetics and modern energy use. Carl De Torres Graphic Design.

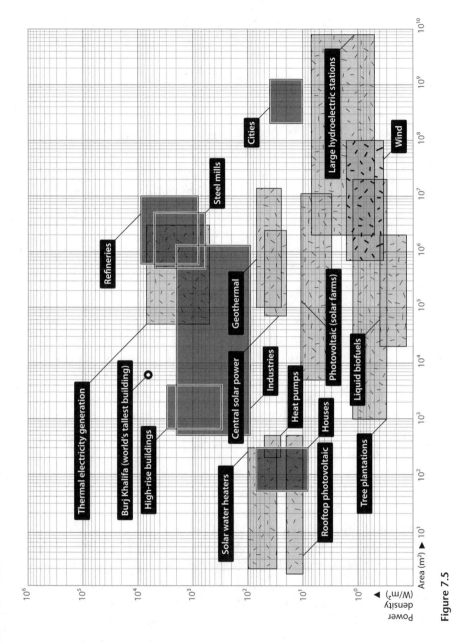

Figure 7.5
Power densities of renewable energetics and modern energy use. Carl De Torres Graphic Design.

two or three orders of magnitude more space to secure the same flux of useful energy if they are to rely on a mixture of biofuels and water, wind, and
solar electricity than they would need with the existing arrangements. This
is primarily due to the fact that conversions of renewable energies harness
recurrent natural energy flows with low power densities, while the production of fossil fuels, which depletes finite resources whose genesis goes back
10^6–10^8 years, proceeds with relatively high power densities This power
density gap between fossil and renewable energies leaves nuclear electricity
generation as the only commercially proven nonfossil high-power-density
alternative. That is why further advances in photovoltaic electricity generation, the renewable conversion with the highest power density, would be
particularly welcome.

Several bold proposals would sever the link between renewable electricity generation and extensive land requirements. They include a variety of
ocean energy conversions—exploiting the kinetic energy of waves and currents and the difference in thermal energy between surface and deep waters
(Charlier and Finkl 2009; Cruz 2008)—and wind generation by turbines
placed within the jet stream (Roberts et al. 2007). None of these proposed
alternatives is likely to evolve fast enough to supply a significant share of
global energy demand (10%–15% of 2013 use would mean 1.7–2.6 TW).
Nor are there any realistic prospects for early, large-scale commercialization
of landless PV conversions using giant buoyant PV panels in the strato-
sphere (StratoSolar 2014) or the Moon-based PV beamed to Earth by micro-
waves (Girish and Aranya 2012).

Aggregate Land Claims

Systematic quantifications of many specific kinds of land use (in addition
to such standard data as arable or forested land) are fairly recent; even such
important categories as urban land or impervious surface area of the United
States have been reliably assessed only during the past 10–15 years. Two
obvious disclaimers: aggregates hide many qualitative differences, and they
have no claims to high accuracy, though the aim is to do better than just
getting the right order of magnitude. I am confident that even for the global
totals, and definitely for the US aggregates, even the most uncertain numbers have error margins no greater than ±50%, and some are off by less
than ±25%.

Global Energy System

Global data for the extraction of fossil fuels, their processing and transportation, and for electricity generation and transmission are fairly reliable (BP 2014; IEA 2014; UN 2014), and I use them, converted to annual powers for the year 2010, together with fairly liberal averages of specific power densities, that is, assuming relatively low rates, in order to err on the high side of aggregate land claims. Moreover, to avoid any appearance of unwarranted accuracy, all itemized results are rounded upward to the nearest 100 km^2, and all category totals are rounded to the nearest 1,000 km^2. The only category that has been deliberately excluded from this global aggregate are the ROWs of railroads that the coal-carrying trains share with other cargo.

The extraction of nearly 14 TW of fossil fuel (including the on-site processing of coal and natural gas) claimed about 12,000 km^2 (and most likely not less than 10,000 and not more than 20,000 km^2), and crude oil refining added about 1,000 km^2. Tanker terminals and natural gas liquefaction facilities occupied only about 300 km^2. With assumptions of either an average ROW width of 15 m or an average throughput power density of 300 W/m^2, the world's refined oil product and natural gas pipelines (whose total length reached about 2 Gm in 2010) preempted other land use on almost 30,000 km^2. Aggregates for global electricity are more error-prone. Fossil-fueled stations required about 1,500 km^2, nuclear power plants added at least 600 km^2. Any errors in these values are negligible compared to the estimates for hydroelectric power generation; my best estimate of land claimed by water reservoirs used for electricity generation is on the order of 100,000–150,000 km^2.

The International Commission on Large Dams' register of dams contains data on nearly 38,000 structures taller than 15 m (ICOLD 2014). About 72% of them are single-purpose dams (50% built for irrigation, 18% for electricity generation). Among the multipurpose dams, irrigation (24%) and flood control (20%) are more important uses than power generation (16%). Consequently, only about 20% of all dams are used solely or primarily to generate electricity, and hence even if we use one of the higher estimates of global reservoir area (about 600,000 km^2), hydroelectric power generation would claim at least 120,000 km^2. ROWs for high-voltage transmission lines can be estimated in two ways: by assuming that the aggregate line length of about 1 Gm claims on average about 5 ha/km, or by assuming that the transmission of 2.3 TW in 2010 proceeded at an average rate of

40 W/m². The two totals are respectively 50,000 km² and about 58,000 km², a close agreement for estimates of this kind of global aggregates.

This summation *(keeping in mind it is just a fair approximation)* means that in 2010 the world's fossil fuel–based energy supply—produced by the extraction of coals and hydrocarbons, their processing and transportation, thermal electricity generation, and the proportional claims of high-voltage transmission ROWs (nearly 70% of the total)—took at least 80,000 and no more than 90,000 km² of the grand total, the latter total being an area smaller than Portugal or Hungary, and with 13.6 TW of primary fossil energy, its average power density (counting the land devoted to fossil fuel–fired electricity generation) was roughly 150–170 W/m².

The energy system that dominated the global supply during the past 50 years—fossil fuels, thermal, and hydroelectricity generation, delivering 14.34 TW in 2010—claimed roughly 230,000 km² (200,000–250,000 km²) of land that is either directly occupied or whose uses are restricted by the ROWs imposed by pipelines or high-voltage transmission lines, which means that it has been operating with an overall power density of about 60 W/m² (fig. 7.6). The mean of 230,000 km² is slightly smaller than Romania; the higher value equals about half of Spain's land surface and is less than 0.2% of the Earth's ice-free land. Leaving pipeline and transmission line ROWs aside, the grand total comes to nearly 150,000 km² (less than half of Poland), of which almost 90% is land flooded by reservoirs.

In 2010, new renewable energy sources (solar PV, wind, liquid biofuels) contributed just 130 GW, or merely 0.9% of all primary commercial energy. PV-based electricity claimed less than 1,000 km², and wind turbines were spread over about 40,000 km² (when assuming an actual generation density of 1 W/m²) and less than 1,000 km² when only the land occupied by turbine pads and associated infrastructures is counted. In 2010 biofuels (dominated by sugar cane and corn grown for conversion to ethanol) supplied only about twice as much power as wind—79 GW versus 40 GW, that is, an equivalent of less than 1.5% of the global crude oil output—but their land claim was more than 260,000 km², an area slightly larger than the UK and equal to almost 2% of arable land planted to annual crops (FAO 2014). Even when calculations were performed using minimal wind turbine claims and excluding all transmission ROWs, modern renewables (excluding hydroelectricity) required almost 270,000 km² to deliver 130 GW, with a mean power density of just 0.5 W/m². They claimed more land than

Box 7.1

Aggregate land claim of the global energy system in 2010

Process	Power	W/m2	~km2
Fossil fuel extraction	13.63 TW		12,000
Coal	4.72 TW	1,000	4,700
Crude oil	5.38 TW	1,000	5,400
Natural gas	3.53 TW	2,000	1,800
Crude oil refining	5.10 TW	5,000	1,000
Fuel transportation			27,000
Hydrocarbon pipelines	8.03 TW	300	27,000
Tanker terminals	2.21 TW	10,000	200
LNG terminals	364 GW	5,000	100
Thermal electricity generation	1.86 TW		2,100
Fossil-fueled electricity	1.54 TW	1,000	1,500
Nuclear plants	316 GW	500	600
Renewable energies	525 GW		398,000
Hydroelectricity	395 GW	3	131,700
Geothermal electricity	8 GW	50	200
Solar electricity	3 GW	5	600
Wind electricity			
Turbine spacing	40 GW	1	40,000
Footprint	40 GW	50	800
Modern biofuels	79 GW	0.3	263,300
Electricity transmission	2.30 TW	30	58,000

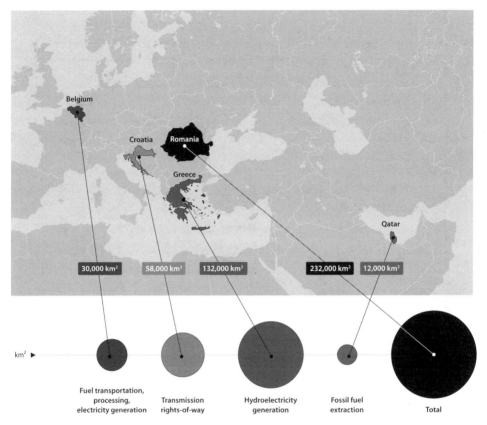

Figure 7.6
Equivalents of land claims of the global fossil fuel–nuclear-hydro energy system in 2010. Carl De Torres Graphic Design.

the global fossil fuel-nuclear-hydro system, which delivered 14.3 TW (110 times as much power) with an average power density of about 60 W/m^2.

Final comparisons are between the power densities of energy production and the final uses of fuels and electricity. The hierarchy of final uses proceeds from annually averaged power densities of around 3 W/m^2 for small, densely populated modern economies to between 10 and 30 W/m^2 for many urban areas to more than 100 W/m^2 for city downtowns. This leads to a number of revealing conclusions. Perhaps most important, the fossil fuel–nuclear-hydro systems that dominate the energy supply in modern affluent economies have been operating with an overall power density that,

Box 7.2
Terrestrial areas modified by human action, circa 2010*

Activity	Area (1,000 km^2)	Percent of ice-free land
Arable land and permanent crops	16,000	12.3
Area affected by logging	3,000	2.3
Forest and tree plantations	3,000	2.3
Urban areas (including roads)	4,000	3.1
Impermeable surfaces	600	0.5
Reservoirs	600	0.5
Fossil fuels extraction	15	0.01
Rights-of-way (pipelines, HV lines)	90	0.07
Hydro reservoirs	150	0.1
Modern energy system	~250	<0.2

Notes: *Data from FAO (2014), Hooke, Martín-Duque, and Pedraza (2012), from previously cited studies of urban areas and impervious surface areas, and from box 7.1.

depending on climate, population density, and level of industrialization, is mostly two to four times that of the power required by large urban areas.

Fossil fuels (when transportation and transmission ROW needs are included) generally supply energy with power densities higher than those prevailing in city downtowns, and the only instances in which the power densities of energy use surpass those of common ways of energy production are the energy-intensive industrial processes (often well above 1,000 W/m^2) and city blocks consisting of densely packed high-rise buildings (on an annual basis they can go well above 500 W/m^2) and during short periods of peak demand (driven by winter heating or summer air conditioning) in downtown cores, where they can go to as much 1,000 W/m^2 or even more.

A table and a graph (fig. 7.7) make it easy to appreciate how small is the absolute claim of the modern energy infrastructure in comparison with other human activities that have modified large parts of roughly 130 Tm^2 of the Earth's ice-free surface:

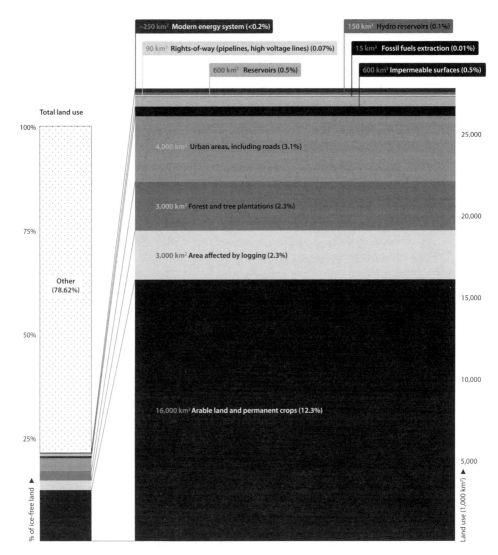

Figure 7.7
Comparisons of global land use in 2010 (land use in 1,000 km²). Carl De Torres
Graphic Design.

US Energy System

The land requirements of America's energy system can be estimated more accurately than the total claimed by the global energy production. I use supply specifics from official statistics (primarily the USEIA) and, once again, assume rather conservative means of power densities, and round the totals to the nearest 50 km^2. In 2010, fossil fuel extraction claimed roughly 6,000 km^2, fuel processing (mainly refining) needed less than 500 km^2, and the ROWs of long-distance hydrocarbon pipelines preempted all but a few concurrent land uses on about 15,000 km^2. In 2010 about 40% of all mass transported by railroads was coal (AAR 2013), but as only a small minority of lines are dedicated solely to coal shipments the only imperfect way to proceed is to attribute roughly 40% of ROWs of high-density "A" track (approximately 3,100 km^2: 30-m ROW for 104,000 km) to coal; that would add about 1,200 km^2. Thermal electricity generation (including the domestic part of the uranium fuel cycle) occupied less than 600 km^2, reservoirs for hydro generation covered at least 17,000 km^2, and transmission ROWs claimed nearly 16,000 km^2.

The entire fossil fuel–nuclear-hydro energy system thus required almost 53,000 km^2 (roughly 0.5% of the US territory, an area roughly half the size of Virginia or Tennessee), of which some 30,000 km^2 (55%) were ROWs and nearly a third water reservoirs, while the fossil fuel–based supply (including pipeline and railway ROWs) claimed only about 19,000 km^2. For the sake of completeness, the nationwide fossil fuel–based total should be enlarged by the area of abandoned wasteland created by the surface extraction of bituminous coal that has yet to be reclaimed. During the first decade of the twenty-first century the difference between new permits for surface coal mining and bond releases (issued after the completion of planned reclamation) was on the order of 250 km^2/year, and that would imply the growth of unreclaimed land debt by more than 2,000 km^2/decade.

But that is both too much and too little: too little because in forested areas (particularly in Appalachian mountaintop removal) even the best reclamation effort cannot recreate the original plant composition, and an ecosystem that would closely resemble it may get reestablished only after decades, even centuries; too much because in some locations (particularly in grassy regions), unreclaimed areas may naturally revegetate in a matter of years without any deliberate reclamation effort. In any case, it should be remembered that the overall land claim attributable to the US coal

Box 7.3
Aggregate land claim of the US energy system in 2010

Process	GW	W/m2	~km2
Fossil fuel extraction	1,847.4		7,550
Coal	7,341	1,000	750
Unreclaimed land			2,000
Crude oil	443.4	100	4,450
Natural gas	669.9	2,000	350
Oil tanker terminals	534.6	10,000	50
Crude oil refining	1,077.6	3,000	350
Fuel transportation		13,000	
Oil pipelines	291,300 km with		4,400
Gas pipelines	15-m ROW		7,400
Coal shipments	104,000 km with		1,200
	30 m ROW		
	40% of 3,100 km^2		
Fossil-fueled electricity	329.1		250
Coal	210.9	1,000	200
Gas and oil	118.2	3,000	50
Nuclear electricity	92.1	300	300
Hydroelectricity	29.7	1.7	17,500
New renewables	46.6		~125,000
Geothermal electricity	1.7	50	50
Wind electricity			
Turbine spacing	10.8	1	10,800
Footprint	10.8	50	200
Ethanol	34.1	0.3	~114,000
Electricity transmission*	470	30	15,700

Note: *Total power includes the burning of wood waste.

extraction is definitely much higher than my approximation, and, choosing to err on the high side, I add 2,000 km^2.

Hydroelectricity is the traditional source of renewable energy, with by far the largest land claim among mature energy conversion, while the relatively small contribution of geothermal electricity (less than 2 GW in 2010) requires less than 50 km^2, and electricity generated by the combustion of woody phytomass (about 5 GW) comes mostly from burning logging residues and does not create any additional space claims and hence is not included in the aggregate count. Of the three kinds of new renewable energy supply, PV-based solar electricity generation was still minuscule in 2010 (accounting for less than 50 MW), but wind turbines contributed 10.8 GW, equal to slightly more than a third of hydroelectric generation. They were spread over an area of nearly 11,000 km^2, but the land actually occupied (mostly by tower pads and access roads) was only on the order of 200 km^2 (less than three Manhattans). But in 2010 nearly 29% of corn and sorghum harvests were used to produce ethanol, and their cultivation claimed about 124,000 km^2, an area larger than Pennsylvania and three times as large as all land (including all ROWs) claimed by the entire US fossil fuel–nuclear–hydro energy system.

Once again, these approximate nationwide summations lead to interesting insights. Coal extraction, thanks to highly productive western surface mines, claims less than 1,000 km^2, and more land is occupied by ROWs of railroad coal transportation when they are apportioned according to coal's share in the annual mass of rail shipments on high-density lines; unreclaimed coal mining land occupies an even larger area that is impossible to estimate with a fair accuracy. Mainly because of a very large number of old, poorly productive oil wells, the average power density of US crude oil extraction is an order of magnitude lower than the country's coal production, and ROWs for oil and gas pipelines are 2.5 times as large as the land taken by the extraction of all hydrocarbons.

The overall power density of fossil fuel production (roughly 1.85 TW, claiming roughly 7,500 km^2) is relatively low, about 250 W/m^2. Imports and the processing of fossil fuels (dominated by shipments of crude oil and refined products and by the refining of domestic and imported crude) make a relatively small spatial claim (less than 500 km^2), a total smaller than an unavoidable margin of error in estimating fuel extraction claims. America's fossil fuel–fired electricity generation (fuels needed to energize it have already been accounted for) occupies less than 300 km², and nuclear generation (including the fuel supply) takes up a similar amount of space—and the land requirements of both of these industries are dwarfed by the land flooded by reservoirs used for hydroelectric production and by ROWs of high-voltage transmission lines.

The country's dominant fossil fuel–nuclear-hydro energy supply system—whose domestic primary power output reached about 2 TW in 2010 and whose land claim was on the order of 55,000 km²—had an overall power density of nearly 40 W/m² as its low-density components (hydroelectric generation and pipeline and transmission ROWs) overwhelmed the high power densities of fossil fuel extraction and processing. Net fossil fuel imports added about 750 GW to the domestic production, and so the power density of the entire system would be about 50 W/m². As expected, the overall power density of the nascent energy supply delivered by new conversions of renewable energy sources is much lower: the growing triad of wind turbine–generated electrocity, solar electricity, and liquid biofuels reached a bit over 60 GW in 2010, and even after counting only the land actually occupied by wind turbines and their infrastructure and excluding all transmission ROWs the new renewable system delivers with an overall power density of just 0.4 W/m², less than 1/100th of the currently dominant arrangements.

Additional annual land claims cannot be estimated by simply applying appropriate power densities to specific expansions of fuel extractions and electricity generation or to an extension of pipelines and high-voltage lines. This is why such simplistic extrapolation would end up with substantial errors: in some localities the land disturbances created by surface coal mining are more than matched by the mandatory reclamation of old abandoned wasteland; directional drilling of multiple wells from a single well pad reduces the specific land claims of new wells; new refinery capacities and new natural gas–fired electricity generation can be accommodated within the sites of existing facilities; and some new transmission lines and some new pipelines can use, fully or in part, existing ROWs. As a result, my best estimate is that recent net annual additions (for land actually transformed by new energy projects and for new ROWs) have been less than 500 km².

I will close this section in the same way as I did the previous one, by contrasting the land claims of US energy production with other important land uses (fig. 7.8). Here they are in descending order: protected areas in

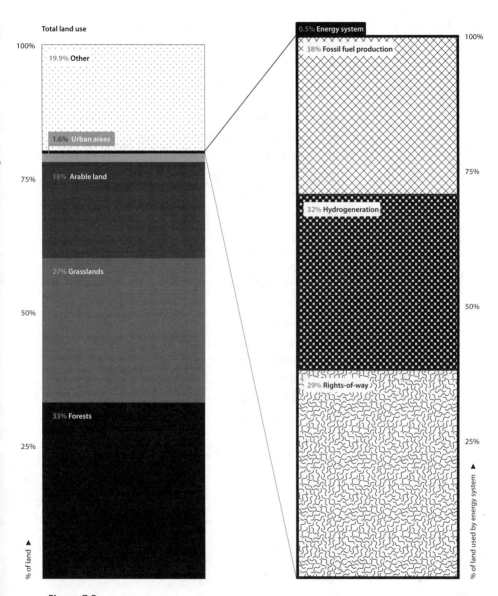

Figure 7.8

Comparisons of US land use in 2010. Carl De Torres Graphic Design.

forested areas are just over 3,250,000 km^2 (33% of the nation), and arable land and permanent crops occupy 1,630,000 km^2 (about 17% of the total), while the recently harvested area of annual crops has averaged less than 1,400,000 km^2. As I have already explained, urban areas (including also grassy and treed surfaces) take up about 150,000 km^2, and impervious surface areas (just buildings and paved surfaces) amount to about 50,000 km^2.

This means that the US fossil fuel–nuclear-hydro energy system (including all ROWs) is slightly more extensive than the country's impervious surfaces, and that the land taken up by cities (including their green surfaces) is roughly three times as large as the dominant energy supply system consisting of fossil fuels and nuclear and hydroelectricity. Inversely, this means that for every 3 m^2 of urban areas (where most of the modern energies are used) there is roughly 1 m^2 of land that is either transformed or whose other uses are largely preempted by the extraction, conversion, transportation, and transmission components of the US fossil fuel–hydro-nuclear energy system.

In contrast, in 2010 ethanol delivered an equivalent of only about 5% of the country's motor gasoline supply but the production of its primary feedstock (grain corn) required roughly four times as much land as the entire US fossil fuel–hydro-nuclear energy system. This brings me to the closing chapter of the book, in which I will look at how much land these new renewables would require if they were to entirely displace our current reliance on fossil fuels and nuclear fission. In other words, I will examine the unfolding energy transition in terms of shifting power densities.

8   Energy Transitions

*Unfolding energy transition changes a system dominated by fossil fuels and ther-
mal electricity generation to new arrangements in which renewable energy conversions become more important. In this chapter I look at this transition through
the prism of power densities and offer some conclusions about the import and
consequences of these new realities.*

We are at the beginning of another epochal shift to new sources of energy.
The timing of the first shift, to using energy other than our muscular exertion, cannot be determined with accuracy as the earliest confirmed dates of
the controlled use of fire have been steadily receding and now appear to be
at least 10^5 years ago (Smil 2013a). The second shift began to take place
more than 7,000 years ago with the domestication of draft animals: their
greater power was a critical component in the evolution of traditional agriculture, land transport, and construction. Only millennia later did the set-
tled societies added the limited use of wind and water power harnessed by
sails, windmills, and waterwheels. Some combination of these traditional
energy sources characterized all preindustrial societies. This pattern began
to change only in the early modern era with the rising combustion of coal
in the UK and in a few European regions; after 1860 hydrocarbons were
added to the rapidly expanding use of coal, and the introduction of internal
combustion engines and thermal electricity generation accelerated the
transition from phytomass to fossil fuels.

Modern civilization is thus a material and intellectual embodiment of
converting fossil fuels into the useful energies of heat, electricity, motion,
and chemical potential. While photosynthesis remains the planet's most
important energy conversion, as it powers all life (with the exception of

deep-sea organisms aggregating near hot thermal vents), our relative reliance on phytomass fuels has been steadily declining. In 1850 coal, the only commercially extracted fossil fuel, supplied less than 10% of the world's primary energy, and during the closing years of the nineteenth century the gross energy content of traditional phytomass fuels (wood and charcoal and straw) and fossil fuels became even.

At the beginning of the twenty-first century, fossil fuels (with crude oil slightly ahead of coal, followed by natural gas) supplied just over 80% of the world's primary energy (including noncommercial biomass burned by poor households), and, because of their more efficient conversions in boilers, furnaces, and engines, they provided more than 90% of all useful energy (Smil 2010b). The worldwide share of fossil fuels in commercial energy supply was almost 87% in 2013, when fossil energies supplied 93% of Japan's, 90% of China's, 89% of Russia's, 86% of the United States', and 83% of Germany's primary energy demand, with the largest share of nonfossil energies originating either in nuclear power (in the United States and Russia) or in hydropower (in China and Japan).

New kinds of renewables—dominated by wind turbines and PV cells in Germany and China and by liquid biofuels in the United States—reached the highest share of the overall supply in Germany, where preferential policies had pushed their share to 9.1% of all primary energy by 2013 (BP 2014). Elsewhere the contributions of new renewables were much lower, amounting to 2.6% in the United States, 2% in Japan, and 1.5% in China, while in Russia (where they accounted for a mere 0.014% of the total supply) these alternatives were essentially absent. These realities make it clear that the transition from fossil fuels to new renewables, much like all previous energy transitions, will be a gradual, protracted affair that will take place over many decades (Smil 2010b).

Two arguments are advanced in favor of an accelerated shift away from fossil fuels: concerns about their future supply and worries about the long-term environmental consequences of their combustion. The first concern has been around for generations but has received more attention recently, thanks to many publications claiming the imminent arrival of peak oil, peak coal, and peak everything (Smil 2010a; 2013b). Production realities do not reflect any convincing signs of an imminent global peak for any fossil fuel extraction, and the best appraisals of remaining coal and hydrocarbon resources indicate a sufficient supply for decades to come: neither the leading government agencies (USDOE 2013a) nor large energy companies (ExxonMobil 2013b) assume otherwise.

The argument can then shift to the quality and cost of fossil fuels. Running out of fossil fuels (of any quality and producible at any cost) may be a slow process with no imminent end, but diminishing reserves in many of the most accessible deposits in the richest coal seams and hydrocarbon reservoirs in the areas that have been explored and exploited for generations are an undeniable reality. One of the latest impressive demonstrations has been the high cost of bringing new crude oil to the market: in 2013 no new major oil project (with at least 300 Mb in estimated lifetime output) added in the previous two years had a breakeven cost below $70 per barrel (Goldman Sachs 2013).

The environmental argument for an accelerated shift away from fossil fuels—the need to slow down anthropogenic global warming—has dominated the quest for new renewables since the late 1980s. The only two technical solutions that would obviate global decarbonization would be the massive development of nuclear power or an equally large-scale effort to capture and sequester CO_2 generated by fossil fuel combustion. Both of these options have their strong advocates, but there is no evidence of any commensurate deployment. Nuclear power is either stalled (United States) or on the way out (EU), with more than 40% of all reactors (66) under construction in China and 75% of them being built in just four countries, China, Russia, India, and South Korea (IAEA 2012; Schneider et al. 2013). And most of the proposed carbon capture and sequestration projects have been shelved or postponed (Global CCS Institute 2014). There is a third option, climate engineering, but its efficacy, practicality, and acceptance are even more uncertain (Keith 2013).

By 2012 nearly 140 countries had policy targets for an energy supply provided by renewable sources, ranging from modest shifts to profound changes: in the UK the target is 15% of all energy from renewable sources by 2020; in Germany it is 30% by 2030 and 80% by 2050 (REN21 2013). Moreover, some studies have argued that it would be perfectly possible to have all new energy demand supplied from renewable sources (wind, water, and solar) by 2030 and the world's total energy demand coming from renewable sources by 2050 (Jacobson and Delucchi 2011). The latest incarnation of this plan sees the state of New York completely energized by nothing but renewably generated electricity by 2030 (Jacobson 2013), a plan found by Dodge (2013) to lack any technical credibility as it assumes unrealistic performance factors and the mass deployment of unproven conversion techniques.

I have shown in a great detail that no worldwide energy transition has ever been—indeed, cannot be—so rapid (Smil 2010b), and that since the rise of coal none has ever resulted in complete domination by a single energy source. Apparently, such realities do not matter where theoretical calculations reign. Of course, a theoretical option might consider only electricity generation: if it were exceptionally inexpensive, that electricity could be used as the foundation of a new hydrogen economy, a solution that has been advocated by some enthusiasts for decades (Ball and Wietschel 2009; Dickson, Ryan, and Smulyan 1977).

In reality, we could not convert every convertible industrial, transportation, residential, and commercial energy use to electricity in just 15 years, and we would still need fuels for iron smelting, for feedstocks (needed to replace fossil hydrocarbons that now dominate many chemical syntheses), and, of course, to energize our land, water, and air transportation: the world's fleets of heavy trucks or container ships appear unlikely to run on electricity or fuel cells anytime soon, and electric or cell-powered commercial jetliners are pure science fiction. That is why any successful transition to new renewable energy sources would have to deliver both electricity and fuels. And, obviously, it would have to do so on multi-gigawatt scales in large countries and on a terawatt scale globally. I will examine first the spatial implications of displacing all liquid fossil fuels by biofuels and all metallurgical coke by charcoal and then take a closer look at wind and solar power electricity in the unfolding energy transition before closing the chapter with quantitative illustrations of what it would take to create an energy system based completely on renewable sources.

Biofuels

Proponents of new renewable energy sources argue that this challenge can be met by relying, as we had for millennia, on phytomass—but doing so by cultivating it with much higher yields, through using a wider variety of plants and converting them to fuels with higher efficiencies. But in in no other instance of displacing fossil fuels by renewable energy sources is there such a mismatch as between the power densities of producing oil-based liquid fuels and the power densities of producing biofuels: the extraction of crude oil and the harvest of phytomass feedstocks differ by at least two, most commonly by three, and in many cases by four orders of magnitude. Moreover, the scale of the needed substitution (roughly 2.5 Gt of transportation liquids in 2012) is enormous: even if we assume no increase or even a slight decrease in the US demand (USEIA 2013a) and restrained growth of the current liquid fuel requirements in Asia, the annual supply of liquid fuels would have to amount to at least 3 Gt (that is, 4 TW) by 2040 (ExxonMobil 2013b).

Many studies have endeavored to show that such mass-scale biofuel futures are possible. When Berndes, Hoogwijk, and van den Broek (2003) reviewed all published long-range forecasts of biomass energy contributions they found that the maxima for the year 2050 ranged from less than 3 TW to nearly 13 TW, the latter total being higher than the world's total primary energy supply (TPES, including traditional biofuels) in the year 2000. Three years later Moreira (2006) claimed that phytomass can eventually supply about 32 TW, which is more than twice as much as all fossil fuels produced in 2012. But all of these high claims rest on overly enthusiastic constructs that assume unrealistically high phytomass yields and exceptional conversion efficiency maxima, posit an abundance of requisite land, and fail to examine a number of other critical constraints. In reality, biofuel production would be constrained by many factors, and the consequences of shifting from fossil fuels to biofuels would be profound. Power densities help to explain some of these outcomes.

I have always been skeptical about any claims of grandiose phytomass futures. In 1983 I concluded my book on biomass energies by writing that "there are countless better uses for plants than to burn them directly or to use them as feedstocks to make fancier fuels out of them" (Smil 1983, 417), and during the past three decades nothing has taken place to make me change that conclusion. Phytomass will continue to be a non-negligible contributor to global energy supply for generations to come, but it cannot provide large fractions (a third, one-half, two-thirds) of that rising total, and differences between theoretical appraisals of potentially available phytomass and realistic opportunities for its use should always be kept in mind.

For example, in 2005 a study commissioned by the US Department of Energy concluded that the country has an annual supply of one billion dry tons (that is, 900 Mt) of agricultural and forest phytomass (USDOE 2005), enough to displace about 30% of oil consumption at that time. But no component (crops, wood, residues, wastes) was restricted by the cost of cultivation and harvesting, and all referred to phytomass on farms or in forests and excluded all transportation and storage costs, handling losses, and quality deterioration. This resulted in a misleadingly huge mass that shrinks on closer examination. The study's update acknowledged those shortcomings and offered a more rigorous assessment (USDOE 2011a). Or, as Sinclair (2009, 407) put it, "while biofuels can be a contributor to the energy needs of the future, realistic assessments of the production challenges and costs ahead impose major limits."

Liquid Biofuels

Crude oil extraction proceeds mostly with densities of 10^2–10^3 W/m², and in the largest fields the rate goes up well into 10^4 W/m²—while the dominant feedstocks for the production of liquid biofuels are harvested with power densities of 10^{-1} W/m² and the rates are further reduced by their processing. A power density of 0.23 W/m², typical for US corn-based ethanol, demonstrates the inherent limits of America's principal alternative automotive fuel. If all of America's gasoline demand in 2012 (a total of 16.96 EJ, or 537.87 GW) were to be supplied by corn-based ethanol produced with that power density, then the United States would have to be growing corn for ethanol on 234 Mha, an area nearly 75% larger than that of all recently cultivated land and a third larger than the country's total cropland (USDA 2013a).

Ethanol's low power density also causes extensive qualitative environmental changes associated with the fuel's production (Howarth and Bringezu 2009; Smil 2010a). Corn cultivation's environmental impacts include soil erosion from the cultivation of a row crop; the demand for heavy applications of nitrogen, averaging more than 150 kg/ha and surpassing 200 kg/ha in the Corn Belt; the ensuing nitrogen leaching, causing the eutrophication of coastal waters and an expanding dead zone in the Gulf of Mexico; the depletion of aquifers for irrigation; the expansion of monocultures as traditional rotations with soybeans and alfalfa; and a net increase in greenhouse gas emissions. In addition, ethanol distilleries discharge large volumes of wastewater (10–13 times the volume of the produced fuel), and those discharges increase oxygen demand in streams and water bodies.

Sugar cane is a superior feedstock: not only has it much higher yields than corn but because of its endophytic bacteria the most productive varieties do not require any nitrogen fertilizer. The crop's 2010 yield averaged nearly 72 t/ha worldwide and 79 t/ha in Brazil. Assuming that the global mean eventually rises to 80 t/ha and that the harvest would be converted to ethanol with Brazilian efficiency (82 L/t), the power density of cane-based ethanol production would be almost exactly 0.5 W/m^2 (6,560 L × 24 MJ/L/31.536 Ms/10^4). Converting the entire global harvest (1.69 Gt in 2010) to ethanol would yield about 105 GW, merely 3% of the global liquid fuel demand in 2010. Production of the world's 2040 liquid fuel demand (4 TW) would require about 800 Mha of the crop, which would take slightly more than half of the global cropland and about 50% more than all the tropical and subtropical farmland suitable for cane cultivation. And supplying only half of all liquid fuel or replacing only all gasoline would still require 300–400 Mha of cane. The only way to get that much land would be to convert more tropical grasslands and forests to cane fields.

But ethanol is not suitable for jetliners, and aviation has been one of the fastest-expanding sectors dependent on high-energy-density kerosene (jet fuel). In 2010 the worldwide demand for kerosene reached about 11.4 EJ, or roughly 360 GW (USEIA 2014a); by 2050 it is expected to reach 1 TW (ICAO 2010). Recent production power densities of biojet fuel range from just 0.06 W/m^2 for a soybean-based substitute to 0.65 W/m^2 for palm oil (Rossillo-Calle et al. 2012). Even the latter alternative would require about 57 Mha of palm oil plantations, 3.5 times their global 2010 area, inevitably leading to a further increase in the tropical deforestation that has accompanied the crop's recent expansion (Kongsager and Reenberg 2012).

Basing the fuel on soybeans would need about 570 Mha of crop dedicated to biojet fuel, 5.5 times the 2010 global total, obviously a highly unlikely (if not impossible) extension. Turning to crops grown on marginal, nonarable land could provide only a partial solution: the much touted jatropha (*Jatropha curcas*, a hardy oilseed-bearing shrub or a small tree able to grow on arid soils) would not produce more than 0.2 W/m^2, and hence the world would need 180 Mha of it to satisfy the 2010 jet fuel demand, and 500 Mha in 2050, an area equal to slightly more than half of China's territory. Even if genetically improved cultivars were to double the yield, the likely jet fuel demand in 2050 would still call for covering roughly an Argentina- or Kazakhstan-sized area with jatropha.

There is clearly room for increasing the yields of previously uncultivated plants (jatropha, switchgrass), but no yield or conversion gains can change the low power densities of the two leading biofuel crops. Global corn yield rose by 65% in the years between 1980 and 2010 and that of sugar cane increased by 30%; hence, even similar increases between 2010 and 2040 would still leave the power densities of corn-based ethanol lower than the recent densities of the US production (2010 yields averaged 5.2 t/ha world-wide and 9.6 t/ha in the United States). In Asia, where corn yields average only 50% of the US mean, the power density of corn-based ethanol would still be below 0.2 W/m^2, while that of cane-based ethanol would remain below 0.7 W/m^2. And while Novozymes' new enzymes promise to increase ethanol yield by up to 5% (Novozymes 2013), average American corn yields have recently declined by much more than that: they peaked at 10.3 t/ha in 2009, were at 9.2 t/ha in 2011, and widespread drought depressed them to just 7.7 t/ha in 2012 (FAO 2014).

Future land claims imposed by high shares of liquid fuels derived from phytomass could be lowered but not massively reduced by producing cel-lulosic ethanol from crop wastes, from plants grown on marginal land, or from surplus phytomass. The annual production of crop residues rivals that of crops themselves: modern cereals have a residue:grain ratio of roughly 1:1, which means that the global output of straw and stover is about 2.5 Gt, and when residues from other crops are added that total rises to about 4 Gt (Smil 2013a). But crop residues are not wastes waiting to be converted to ethanol: in traditional agricultural societies they are still important sources of cooking fuel and animal feed, and their recycling is a key ingredient of proper agroecosystemic management as straws, stalks, and leaves return to soil nitrogen, phosphorus, and potassium, as well as many micronutrients; renew soil organic matter; help to retain moisture; and prevent wind and water erosion (Smil 1999, 2013a).

This means that only a carefully assessed fraction of crop residues should be removed. Such a restriction would further lower the inherently low power densities of their production: these are (with a 50% removal rate) less than 0.08 W/m^2 for Great Plains winter wheat, about 0.3 W/m^2 for high-yielding German wheat, and (with a nearly complete removal) less than 0.5 W/m^2 for sugar cane bagasse. Roughly a third of stover (the residue of corn, America's largest crop) can be removed in conventional cropping, 70% with no-till cultivation; and a weighted mean of 40% translates to an annual harvest of some 80 Mt in terms of dry weight (Kadam and McMillan 2003).

That would produce between 20 and 25 GL of ethanol, or just 3% of the US gasoline supply in 2010, and it would imply a final product power density of only 0.06 W/m^2. As for any surplus woody phytomass, a review by Smeets and Faaij (2007) offers a corrective quantitative perspective. While they estimated that the world's theoretically available surplus wood (after satisfying the demand for traditional fuelwood and timber) is about 71 EJ (6.1 Gm^3), the technical potential is 64 EJ, economic considerations reduce it to 15 EJ, and the inclusion of ecological criteria nearly halves it, to 8 EJ, or to only about 250 GW, an equivalent of less than 2% of the world's 2010 supply of fossil fuels.

Most important, further adjustments of previously cited power densities are necessary to take into account the energy costs of crop cultivation and fuel production. Dijkman and Benders (2010) calculated the net power densities (they expressed them in GJ/ha/year) for actual bioethanol production from sugar beets and biodiesel production from rapeseed for three specific European cases. The lowest net power density, for Spanish bioethanol, was just 0.02 W/m^2; the highest rate, for Dutch biodiesel, was 0.08 W/m^2. These low rates mean that the replacement of crude oil–derived liquid fuels in Europe would require (even if the mean value rose to 0.1 W/m^2) feedstocks grown on an area of more than 600 Mha, six times larger than all arable land in EU-27 (FAO 2014).

Similar order-of-magnitude calculations are easy to make for replacing fossil fuels by any kind (liquid or solid) of phytomass fuels. The world's 2012 fossil fuel consumption reached roughly 10.85 Gt of oil equivalent (BP 2014). That is (at 42 GJ/t) 455 EJ, or 14.45 TW, and if we assume that the 2050 demand will be just 40% higher (many forecasts indicate a much larger expansion of demand), then 20 TW are candidates for replacement. If only half of that rate were supplied by biofuels (with the total split between the combustion of woody phytomass to generate electricity and the conversion of phytomass to liquids or gases), then each of these biofuels would have to provide about 5 TW. Even if these conversions could be done with fairly high efficiencies—0.5 W/m^2 for woody phytomass and 0.3 W/m^2 for field crops—then crops for energy would have to be harvested from nearly 1.7 Gha of arable land, and wood for energy would require annual harvests of trees from 1 Gha.

But that would mean that in 2050 the area of energy crops would have to be nearly 10% larger than the total area of arable land and permanent plantations in 2012, and the area of continuously harvested tree plantings would be equal to nearly 30% of all of today's closed forests (with canopies covering more than 40% of the ground). This would come on top of the higher future demand for food and wood and in a world of diminishing biodiversity. And yet some uncritical promoters of phytomass energies see no problems with this as they keep conjuring large areas of uncultivated land or assume that more forests and grasslands can be converted to crop-lands. Read (2008) envisioned having an additional 2.4 Gha of rain-fed arable land (compared to the 2013 total of about 1.55 Gha), mostly in the tropics. Marland and Obersteiner (2008, 335) concluded that "it is not now clear if this vision is a dream or a nightmare"—but I have no problem seeing it as truly nightmarish.

We are already harvesting a significant share (close to a fifth) of the biosphere's annual productivity (Smil 2013a), and any future massive increase in biofuel production would have to increase this intervention and would result not only in competition with food and timber production but also in the further weakening of environmental services, particularly those provided by mature forests. And perhaps the greatest irony is that those who would claim to displace fossil fuels by phytomass in order to reduce carbon emissions are apparently unaware that the expansion of land devoted to biofuels can bring the very opposite outcome.

Fargione and co-workers (2008) showed that converting rain forests, peatlands, and grasslands in order to cultivate food crop–based biofuels in Brazil, Southeast Asia, and the United States could release 17–420 times (one to two orders of magnitude) more CO_2 into the atmosphere than the annual greenhouse gas reductions resulting from the displacement of fossil fuels by these cultivated biofuels. Similarly, Searchinger and co-workers (2008) demonstrated that corn-based ethanol does not produce, as previously claimed, substantial CO_2 savings; rather, its production nearly doubles CO_2 emissions over thirty years and increases greenhouse gases for 167 years, while biofuels from often highly touted switchgrass increase emissions by 50%.

Many problems would accompany any large-scale phytomass harvesting for energy, among them the further destruction of natural ecosystems, demands for nutrients and water, the extension of monocultures, vulnerability to pests and diseases, and competition with land uses to grow food and feed. Expanded deforestation would be among the most likely consequences of any global-scale push for much increased biofuel production, and yet even without such pressures the world's forests have been in retreat: the best high-resolution global mapping showed that between 2000 and 2012 there was a net forest loss of 1.5 M km^2, with rising losses in the tropics (Hansen et al. 2013).

As I have demonstrated, the inherently low efficiency of phytomass production and subsequent energy losses arising from various fuel conversions limit the power densities of final forms of phytomass-based energy use to a fraction of 1 W/m^2. This is the key reason why it is most unlikely that modern societies will soon come full circle and return from fossil fuels to phytomass fuels as the dominant source of their energy, and why any responsibly handled expansion of phytomass production and its least wasteful conversions will be able to make a non-negligible but inherently limited contribution to the world's primary energy supply.

Metallurgical Charcoal

As promised in the introductory chapter, I will now assess a surprisingly neglected aspect of energy transition from fossil to renewable sources, the replacement of coke in iron smelting by charcoal from tree plantations. This requires closer looks at modern blast furnaces and at prevailing plantation yields and charcoaling methods. The charcoal supply would have to energize the annual smelting of just a bit over 1 Gt of iron (the average rate for the years 2008–2012). I will use assumptions based on the best plausible case, Brazilian charcoal-based iron smelting, with recent data on eucalyptus plantation yields, charcoaling efficiencies, and blast furnace charges from Sampaio (2005), Swami and co-workers (2009), Peláez-Samaniego and co-workers (2008), Piketty and co-workers (2009), Pereira and co-workers (2012), and Pfeifer, Sousa, and Silva (2012).

In 2010 recycled metal accounted for about 25% of the global steel output, and only about 70 Mt came from the direct reduction of concentrated ores using natural gas. Iron smelted in charcoal-fueled blast furnaces has accounted for only about 0.5% of the total (Sampaio 2005), while iron from coke-fueled blast furnace iron reached 1.03 Gt in 2010 and surpassed 1.1 Gt in 2012 (World Steel Association 2013). The world's largest furnaces now have volumes in excess of 5,000 m³: Shougang Jing Tang's furnace in Caofedian (blown in 2009) has a volume of 5,500 m³, ThyssenKrupp's Schwelgern 2 (since 1993) has a volume of 5,513 m³, and Japan's Japan's Oita 2 was enlarged to handle a volume of 5,775 m³ in 2004 (Hoffmann 2012; Smil 2008; ThyssenKrupp 2012).

Each of these furnaces requires more than 2.5 Mt of coal equivalent to energize a daily output of more than 10,000 t of hot metal. I specify coal equivalent rather than coke, because in modern iron-making coke is partially substituted by coal dust, fuel oil, or natural gas blown directly into a furnace, and even by peletized plastic waste. The entire smelting operation has become much more energy-efficient, and specific energy requirements for coke (t coke/t of hot metal) have been steadily declining. The typical consumption of dry coke per tonne of hot metal declined from about 1 t in 1950 to just 0.6 t in 2000, and the best operations needed only about 450 kg of coke/kg of pig iron (de Beer, Worrell, and Blok 1998; Smil 2008).

At the same time, the global blast-furnace output has been increasing, from less than 50 Mt in 1900 to 580 Mt in 2000, and then, mainly as a result of China's production surge, to 1.035 Gt in 2010, and the global demand for coke has reached record levels: in 2010, 900 Mt of coking coal were converted to about 650 Mt of coke. Hydrocarbons and coal dust directly injected into blast furnaces were equivalent to roughly another 100 Mt of coke, resulting in an annual energy input (30 GJ/t of coke) on the order of 22 EJ of fossil fuels. Metallurgical coke energizes the high-temperature melt (1,300–1,600°C) and acts as the reducing agent: its combustion generates CO_2 ($C + O_2 \rightarrow CO_2$), whose reduction yields CO ($CO_2 + C \rightarrow 2CO$), and that gas reduces oxides into elemental iron ($Fe_2O_3 + 3CO \rightarrow 2Fe + 3CO_2$). Coke also provides support for the heavy charge of iron ore and limestone (added to remove impurities), but it is sufficiently permeable to allow the ascent of reducing gases. In a world run solely by renewable energies, we would have to go back and replace all of these fuels by charcoal made from woody biomass.

With very similar charcoal and coke energy densities, straightforward replacement of the energy used in primary iron smelting would have called for approximately 750 Mt of charcoal in 2010. But charcoal is a much softer material than coke: depending on the wood species, its compressive strength varies between 10 and 50 kg/cm^2, compared to 130–160 kg/cm^2 for coke, and hence charcoal could not support the massive iron ore and limestone charges without getting crushed. The desirable bulk density of metallurgical charcoal is at least 0.4 g/cm^3, charcoal from eucalyptus has 0.53–0.59 g/cm^3 (Pereira et al. 2012), but many wood species yield charcoal densities of only 0.28–0.4 g/cm^3.

That is why Brazilian iron-makers cannot replace all coke by charcoal: economies of scale, competition with foreign producers, and basic technical considerations (charging large furnaces with friable charcoal would cause serious equipment damage) make that impossible (NCIB 2012). As a result, the internal volume of modern charcoal-fueled Brazilian blast furnaces is an order of magnitude smaller than that of the coke-fired units deployed around the world: the biggest Brazilian furnace has just 568 m^3 (Pfeifer, Sousa, and Silva 2012). The construction of larger numbers of smaller furnaces (and the ensuing acceptance of higher metal prices) would not change the necessity of harvesting large amounts of woody phytomass. We can calculate the resulting wood demand based on Brazilian commercial practices.

Most of Brazil's steel comes from modern enterprises that integrate coke-based smelting of pig iron with the production of continuously cast steel, but about a third comes from operations that use charcoal in small blast furnaces concentrated in the states of Pará, Minas Gerais, and Mato Grosso do Sul (Uhlig 2011). An increasing share of charcoal is made from eucalyptus, whose plantations covered nearly 5 Mha in 2011 (Pereira et al. 2012)— but as much as third of the total charge may be much cheaper wood that comes from the illegal cutting of natural forests (Monteiro 2006). Uhlig (2011) estimated that as much as 15% of the Amazon's deforestation could be ascribed to charcoaling. These shares are uncertain because of significant disparities in nationwide estimates of the country's charcoal production (Ghilardi and Steierer 2011).

About 80% of the fuel is made in small brick-and-mud semicircular kilns commonly called *rabo quente*, hot-tail (fig. 8.1). These 2.5-m-high beehive structures, usually grouped in rows by dozens, are stacked with air-dried wood, set alight, and let to smolder for five to seven days; three days after they are extinguished, men enter the kilns and unload the charcoal. The working conditions in many of these charcoaling operations have been

Figure 8.1
Brazilian eucalyptus plantation. International Forestry Resources and Institutions
(IFRI 2012).

described as akin to slave labor that is also highly hazardous (Greenpeace
2013). Workers removing charcoal from ovens are exposed to high temperatures, dust, and smoke; there is long-term exposure to uncontrolled
emissions of nitrogen and sulfur oxides, benzene, methanol, phenols,
naphthalene, and polycyclic aromatic hydrocarbons (Kato et al. 2005). In
mass terms these small kilns convert only between 22% and 27% of wood
into charcoal, and the average nationwide rate is no more than 25% (Bailis
et al. 2013; Swami et al. 2009).

An average blast furnace charge of 450 kg C/kg of pig iron could be supplied by about 500 kg of coke or 630 kg of charcoal (Sampaio 2005). The
minimum theoretical requirement would thus be about 650 Mt of charcoal,
which means that the global rate of charcoal production would have to
increase at least 14-fold compared to the year 2010, when the FAO put its
output at 47 Mt (FAO 2014). With an average charcoaling efficiency of 25%,
the global iron smelting at the 2010 rate of 1.035 Gt would annually

consume 2.6 Gt of wood. Another (official) Brazilian source gives average rates of 2.2 m^3 of charcoal and 4.4 m^3 of wood per tonne of pig iron (Secretaria de Estado de Meio Ambiente do Pará 2008). With a mean eucalyptus wood density of 0.55 t/m^3, that translates into 2.42 t of wood per tonne of pig iron and 2.5 Gt of roundwood that would be needed to smelt the world's 2010 iron production.

The real total would have to be higher because of the transportation and handling losses that would inevitably arise during the export of tropical charcoal to temperate climates, particularly the fuel made from less dense woods than eucalyptus clones. An alternative would be to transport entire logs and set up massive charcoaling facilities in high-income countries, but that would require an unprecedented trade in wood. In 2010, all wood traded globally amounted to about 170 Mt (FAO 2014); exporting most of the wood needed to make charcoal would mean an order of magnitude increase in shipments.

And how does the total of 2.5 Gt of charcoaling wood (or, with a 10% markup for losses, 2.75 Gt) compare to today's global roundwood harvest? According to the FAO, that total wood harvest reached 3.4 Gm^3 in 2010 or (assuming 0.65 t/m^3) about 2.2 Gt (FAO 2014), and a generous addition of 15% for illegal logging would bring it to 2.5 Gt. Consequently, a global charcoal-based iron smelting that would replicate Brazilian practices would alone claim as much (or slightly more) wood than the world's total (legal and illegal) 2010 wood harvest for fuel, lumber, and pulp. Another way to look at this is that in 2010 (assuming an average 25% conversion efficiency), less than 10% of the world's wood harvest was converted to charcoal, while charcoal-based iron smelting at the 2010 level would require converting all of the world's harvest of wood into charcoal, leaving nothing for other uses—and would require a doubling of recent wood harvests to keep the world supplied with timber, veneer, plywood, fuel, and paper pulp.

The extent of the harvested area would depend on prevailing yields, and those, in turn, vary with wood species, soils, climate, and plantation subsidies. With an annual mean increment of 10 t/ha, the global pig iron output in the year 2010 would have required harvests of at least 250 Mha (2.5 Tm^2), an area equal to 60% of Brazil's Amazon rain forest of 4.1 Tm^2 or to nearly half of the total forested area of 5.5 Tm^2 in the entire Amazon basin. With 19 MJ/kg of dry wood—the average value for eucalyptus clones, according to Pereira and co-workers (2012)—that would imply an average power density of 0.6 W/m². A large-scale transition to charcoal-based iron smelting would lead to many improvements, and hence it is realistic to make more optimistic assumptions.

For example, Piketty and co-workers (2009), assuming an average (dry matter) wood yield of 16 t/ha and a 30% carbonization yield (for 80% C content with 10% handling loss), put the land requirement at 1,290 km² for 1 Mt of hot metal, and hence the global iron output fully energized by charcoal would require roughly 130 Mha of high-yielding tropical plantations. Even better performance is conceivable. Modern continuous charcoaling methods (retorts) should have average conversion yields of 35%–40% (Rousset et al. 2011), and cultivation of high-yielding clones in eucalyptus plantations should bring annual wood increments as high as 25 t/ha (Pfeiffer, Sousa, and Silva 2012). In combination, those increments would boost the charcoal yields per hectare fourfold when compared to my initial assumptions, and the global iron smelting at the 2010 level could be supported by less than 70 Mha of tropical plantations.

But all of these calculations are questionable because even the existing Brazilian plantations have serious environmental impacts, because it is most unlikely that all charcoal would come just from Brazil, and because the global iron smelting will continue to increase in order to supply the enormous demand for steel, mainly in Asia and Africa. Extensive areas of tropical eucalyptus (and pine) plantations have been already described by Brazilian environmentalists as *deserto verde*, destroying biodiversity and maintaining monocultures through intensive applications of herbicides (Reportér Brasil 2011). A common practice in Brazil is to harvest trees in a five-year rotation and coppiced for three cycles: after 15 years of growth, glyphosate is applied to kill the remaining rootstock, and new seedlings are planted (Bailis et al. 2013).

Monocultural plantations also increase the rate of soil erosion, divert water from nearby farms, and contaminate runoff with agrochemicals, and their expansion would compete even more extensively for land that would otherwise be used for crops or pasture. In any case, it is extremely unrealistic to expect that all charcoal needed for global pig iron smelting could be made only from harvesting intensive plantations of Brazilian eucalyptus hybrids—and if a significant share of charcoal were to come from temperate species, then the average wood yields would be substantially lower. The short-term cultivation of small plots has impressively high yields, but studies that look at productivities in various environments most likely to be used for tree cultivation (abandoned farmland, former forest land) show a predictably wide range of outcomes.

Truax and co-workers (2012) found that for the productivity of hybrid poplar plantations on abandoned farmland in Quebec, the site effect (elevation, climate, soil fertility) was far more important than the clone effect, with annual yields as high as 22.4 m³/ha (about 11 t/ha) in bottomlands and as low as 1.1 m³/ha (about 0.5 t/ha) on the poorer soils of hill slopes. In northern Italy, Paris and co-workers (2011) harvested 15–20 t/ha/year, but only with fairly heavy nitrogen applications (300 kg N/ha), while achieving means between 10 and 14 t/ha elsewhere, while in southern Italy (Latium and Molise), Di Matteo and co-workers (2012) recorded annual poplar yields of 10–13 t/ha. And a review of 21 studies of poplar and willow plantations in Europe and the United States found yields between 5 and 16.8 t/ha (Djomo et al. 2011). Consequently, assuming a future average global wood increment of 15 t/ha would not be too conservative.

The global output of pig iron doubled between 1990 and 2010, and even if it were to grow only half as fast in the next two decades it would reach about 1.6 Gt in the year 2030. Using the more optimistic assumptions regarding future wood and charcoal productivities it would thus require at least 2.9 Gt of charcoal and annual harvest of wood from more than 190 Mha in 2030. This would be an area of forest or tree plantations only slightly smaller than in the first scenario, and equal to nearly half of Brazil's Amazon rain forest or to more than China's total forest land. And this massive spatial claim would have enormous environmental repercussions, particularly when most of this wood would have to be grown in high-yielding monocultural plantations that would require large inputs of fertilizer, pesticides, herbicides, and, in many drier climates, at least supplementary irrigation.

The world whose dominant metal would be smelted with charcoal produced by the annual logging of an area equal to half of the Brazilian Amazon is conceivable, but it is hardly desirable, and we simply cannot appreciate all of its eventual consequences. But a fundamental concern is clear: could such massive tree harvests (required to produce a billion tonnes of charcoal a year) offer a truly renewable alternative, given their impacts on biodiversity and soil erosion and their constant requirements for water, nutrients, and protection against pests? What would be the real cost of this enormous enterprise, and how practical would it be even if the costs were a secondary matter?

Even if it turned out to be more practical than we think today, the industry based on harvesting wood on a semicontinental scale would be a very different one compared to our current arrangements, in which iron production is based on carbon-rich coke made from coal that is extracted in just a few thousand large mines in a dozen major coal-producing countries and that is easily distributed to large industrial centers for coking. Of course, our current practice taps a finite energy resource, but in light of its importance for iron smelting, we could accord it a highly preferential status and keep relying on it for many generations to come. That would not be difficult to do because we could replace coal's largest use, for electricity generation, with other energy sources.

Wind and Solar Electricity

Both of these flows differ fundamentally from biofuel harvests and conversions. On a positive side, they have substantially higher power densities, and wind turbines can share other productive (agricultural, pastoral, silvicultural) activities in the area required for their optimized spacing. But staggered wood or crop harvesting and phytomass storage (entire trees, wood chips, baled straw) can ensure a continuous supply, while intermittent radiation and wind have, at best, moderately high and often fairly low capacity factors.

Wind

Conversion of wind to electricity has been undoubtedly helped by its light footprint, as the surfaces actually occupied by wind turbine foundations, access roads, and transmission towers amount to a small fraction of the area that contains properly spaced machines. In acknowledgment of this reality I have not used the power densities of a wind farm when quantifying the aggregate claims of energy systems but have counted only actual footprints approximated by a high power density of 50 W/m^2.

And yet using this apparently rational choice is misleading because the power densities of machine spacing are highly relevant for the expansion of the industry. In the United States, where relatively fast and fairly persistent winds prevail across large parts of the Great Plains between northern Texas and North Dakota, there is no imminent prospect of running out of windy sites, but if the wind projects in the region keep expanding they will run into a fundamental power density imperative, namely, the limits of power production by wind farms larger than about 100 km^2. As shown by Adams and Keith (2013), wind turbine drag on local winds would limit electricity generation by these large installations to no more than 1 W/m^2.

Germany faces a different problem. The earliest stages of wind power development took place in localities that combined high wind speeds with persistent flows, that is, in the coastal regions of the windy northern states of Niedersachsen and Schleswig-Holstein. This first phase was to be followed by massive offshore projects, but the costs of such projects—and the costs of and delays in connecting them to land by undersea cables, and then transmitting electricity to the south—refocused attention on the country's central and southern regions, to sites with lower wind speeds and lower wind frequencies. As a result, there are now plans to build 60,000 new wind turbines in orchards, vineyards, and forests (which will require extensive tree felling), on mountaintops (which will require new access roads for the trucks and heavy cranes used to transport and assemble turbines), and even in protected areas in states ranging from Nordrhein-Westphalen to Bayern and from Baden-Württemberg to Sachsen (Schulz 2013).

Such an expansion would greatly alter the appearance of German landscapes, particularly with the invasion of many forested areas, a dubious quest in light of the inherently low-capacity factors in many of these central and southern locations. Still, to meet the country's ambitious goal of 35% of energy coming from renewable sources by 2020, many proponents of massive wind power development offer no relief. Winfried Kretschmann, minister-president of the state of Baden-Württemberg and the first Green Party president of the German Bundesrat (in 2012–2013), insists (as quoted in Schulz 2013) that "es führt kein Weg daran vorbei, die Landschaft auf diese Weise zu verschandeln" (there is simply no other way but to disfigure the countryside like this).

How much this will be opposed in Germany remains to be seen, but the UK has already seen a great deal of opposition to what is perceived as defacement of the country's remarkable landscapes. Wind farms "desecrate our national heritage" (McMahon 2011, 18), "as if some malevolent creature from mythology shed its spawn over the land" (Etherington 2009, 10). Again, the insult would be more tolerable if much higher power densities of wind machines allowed concentrating electricity generation into a much smaller number of locations. And disfiguration of landscapes is not the only consequence of the limited power densities of wind power.

The need to install large numbers of machines tends to reduce the width of noise exclusion corridors, to increase the chances of large-scale bird fatalities, and to affect many terrestrial species as a consequence of the fragmentation of their habitat. In windy and sparsely inhabited Scotland, the rule is to allow 2 km between wind farms and the edge of cities and villages, while in densely populated German states the minimum distances from houses are just 500 (in Bayern), even 300 m (in Sachsen). And, obviously, large wind farms with hundreds of smaller machines in forested and mountainous terrain will kill more birds—particularly raptors, which tend to use mountain slopes and ridge saddles between hills (Subramanian 2012)—than single machines or a small grouping of large wind turbines located on flatlands.

PV Generation

Germany is also the prime example of forcing (through high subsidies) massive installations of PV panels in the country where solar electricity generation has inherently low power densities not only because of the relatively high latitudes (roughly 47°–55°N) but also because of very low capacity factors in climates governed by the prevailing frontal flows from the Atlantic: in 2012 the country had nearly 40% more PV capacity (32.4 GW) than sunny Spain, Italy, and Greece combined (23.7 GW). But in 2013 Germany's Siemens, one of the key benefactors of the country's solar boom, published a study whose conclusion was that building and expanding Europe's solar and wind installations in wrong locations was costing €45 billion (about $60 billion) in unnecessary investment (Siemens 2013b).

I read this with astonishment: one of the world's largest engineering firms had apparently discovered the power of power densities! Details are telling: if Europe's new PV capacities that are to be built by 2030 (about 140 GW) were located in the sunniest Mediterranean sites, the EU could save 39 GW of installed capacity, even after accounting for the cost of additional south-to-north transmission. At the same time, Siemens and several other major German companies (since 2009 grouped into Dii GmbH with other EU partners) was a member of a consortium pushing for a solution that is the very opposite of PV expansion in Germany. The DESERTEC

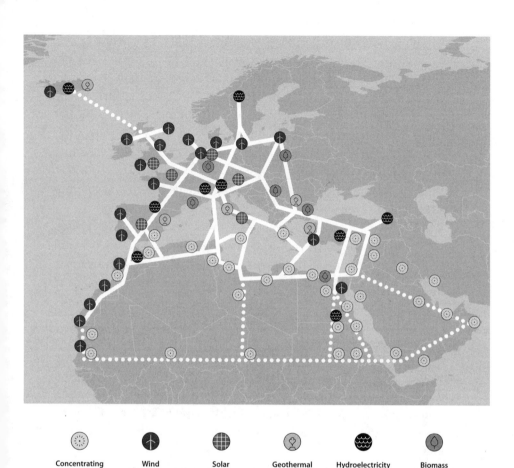

Concentrating solar power — Wind — Solar photovoltaics — Geothermal — Hydroelectricity — Biomass

Figure 8.2
DESERTEC. Carl De Torres Graphic Design. Based on DESERTEC 2014.

Project (or EUMENA—EU and the Middle East and North Africa) is to make PV-based electricity generated in Sahara and the Middle East the principal source of European demand (DESERTEC 2014; fig. 8.2). The scheme would take advantage of the highest possible power densities and would rely on concentrated solar power (CSP), rather than PV cells, to provide a more evenly distributed supply.

Post-2010 upheavals in North Africa and the Middle East have accentuated the project's major weakness: costs and the need to construct high-voltage links of unprecedented capacity aside, would it be wise to put large

CSP facilities in Morocco, Algeria, Tunisia, Libya, Egypt, Syria, Saudi Arabia, and Iraq when most of these countries are not only politically unstable but also have been subject to violent conflicts, and then to transmit all electricity through a few choke points (Gibraltar, Sicily, Bosporus)? In this case the highest possible solar power densities are easily trumped by the inherent riskiness of the endeavor. In any case, Siemens (and Bosch) left the consortium in 2012, and there are no imminent prospects for DESERTEC becoming a major source of EU electricity. DESERTEC also highlights the challenges of servicing very large areas of solar collectors or reflectors in arid environments full of airborne dust.

Regular cleaning (at least two to three times a year) of these exposed surfaces with power washers or brushes would obviously be both water- and labor-intensive, and in many locations the cost of water would surpass the cost of electricity. The best solution appears to be the deployment of robots that glide along rows of PV panels, an innovation by Greenbotics of California that also brings large water savings (SunPower 2013). But this option requires higher inputs of energy-intensive materials to make the devices and more electricity to operate them, thus reducing the net power generation of a solar facility. Robotic cleaning might be a good solution for China with its extraordinarily high air pollution and chronic shortages of water, but more would be required for the DESERTEC projects, where many near-ground structures could be buried under sands in a matter of months.

Another complication whose true dimension cannot be fully answered is the longevity of PV panels mounted on rooftops. They make no new spatial claims as they are merely additions to built-up (impervious) areas, but the length of their service may be rather short in many rapidly growing Asian cities, where their optimal placement may also be fairly limited owing to common shading and where their performance could be heavily degraded by extraordinarily high levels of air pollution (with airborne particulate matter in China's largest cities reaching maxima an order of magnitude higher than the WHO air quality standards).

Finally, a few clarifications are in order regarding the often cited claims that imply extraordinarily high power densities of solar generation. Calculations that divide national electricity demand by peak solar power in sunny regions are quite misleading. For the United States, the 2010 electricity demand of 422 GW could be supplied by an area of some 2,800 km^2

(a square with sides of 53 km) in Arizona—but only if that surface could generate year-round, with a noontime peak power density of 150 W_e/m^2. The National Renewable Energy Laboratory offers a somewhat less misleading number when it claims that "a 100-by-100 mile area of Nevada could supply the United States with all of its electricity" (NREL 2013, 1). In reality, that area (25,600 km²) could—with an average insolation of 220 W/m^2 (Las Vegas mean), a capacity factor of 25%, and a high average conversion efficiency of 15%, and hence with an average annual power density of 8.25 W_e/m^2—generate about 211 GW, or half of the 2010 US demand.

Realistic theoretical estimates are highly dependent on the assumptions used to construct a massive imaginary nationwide PV system. Denholm and Margolis (2008a) used long-term data on average irradiation for the 48 contiguous states assumed a combination of 25% of rooftop and 75% ground-based modules (40% fixed arrays with a 25-degree tilt, 25% single-axis and 10% two-axis tracking) coupled with appropriate long-term storage and ended up with total requirements of 181 m²/capita—compared to per capita averages of 35 m² for airports (or golf courses), 65 m² for roofs, 162 m² for major roads, and nearly 840 m² for urban areas. This means that in absolute terms, entirely PV-based US electricity generation would require about 55,000 km², slightly larger than the combined area of Massachusetts and Vermont and about 0.7% of the total area of the 48 states, and the average power density would be a realistic rate of less than 8 W_e/m^2.

In (largely) gloomy Germany the area needed by PV panels to supply all electricity generation (nearly 560 TWh in 2012) would be considerably larger. With an average PV output of 100 kWh/m² (the recent annual mean for both roof- and ground-based installations), it would require about 5,600 km² covered with modules. That would be the equivalent of nearly 1.6% of Germany's total area, 25% of the country's built-up area, or almost 15% of land claimed by settlements and transportation infrastructure; and roughly 2.7 times the total area of all German roofs, based on an estimate of roughly 25 m² of roof area per person (Waffenschmidt 2008).

What It Would Take

If you are willing to engage in unbounded science and engineering fiction, then, according to Jacobson and Delucchi (2011), this is what it would take to supply the world with 100% renewable energy in 2030 by using electricity (generated by wind, water, and solar PV installations) and electrolytic hydrogen for all purposes: 3.8 million 5-MW wind turbines, 49,000 300-MW central solar plants, 40,000 300-MW solar PV plants, 1.7 billion 3-kW rooftop PV installations, 5,350 100-MW geothermal plants, 270 new 1.3-GW hydro stations, 720,000 0.75-MW wave devices, and 490,000 1-MW tidal turbines. All of that would require only about 0.4% of the world's land for its footprint and 0.6% for spacing, and we are assured that the "barriers to the plan are primarily social and political, not technological or economic" as the energy cost in a new wind-water-solar world "should be similar to that today."

These assurances asides, the simplest reality check shows the fictional nature of these assumptions. In 2013 the worldwide capacity in wind turbines reached about 330 GW, while 13 TW (40 times as much) would be needed by 2030. Total rooftop and large plant PV capacity reached about 100 GW, but 17.1 TW of these installations would be required (170 times as much); moreover, there was not a single 300-MW solar PV plant (five plants rated between 200 and 250 MW), whereas 40,000 would be needed by 2030. In 2013 there was only one central solar power facility rated at more than 300 MW, Ivanpah, at 392 MW, but nearly 50,000 such facilities would be needed by 2030 (an increase of four orders of magnitude). There were fewer than 50 geothermal stations rated at more than 100 MW, but 5,350 would be needed (a 100-fold increase). Pelamis (2014, the world's most advanced wave energy company, produced six 0.75-MW devices by the beginning of 2014, but 720,000 would have to be operating by 2030 (an increase of five orders of magnitude). Finally, by 2013 there were fewer than ten small tidal stations with aggregate installed power of much less than 1 GW, while 490 GW would have to generate by 2030 (two orders of magnitude more).

Such a ramping-up of all kinds of capacities—design, permitting, financing, engineering, construction, all going up between one and five orders of magnitude in less than two decades—is far, far beyond anything that has been witnessed in more than a century of developing modern energy systems. And that still leaves out two other key facts, namely, that such a gargantuan renewable energy system would need an enormous expansion of high-voltage transmission and would require the creation of an entirely new, hydrogen-based society. I still am not sure how we would fly with hydrogen (or electricity) or smelt pig iron. In any case, the chances of a

100% water-wind-solar world to be ready by 2030 are nil, but it is worth while exploring what it would (realistically) take to create an increasingly nonfossil global energy system.

Making Choices

To begin with, even though solar generation has the highest power densities (which may further improve with more efficient PV modules), it would be a very unwise engineering choice to aim at 100% PV-based electricity generation even if moderately good storage options were available. Any sensibly designed all-renewable system would aim at combining different techniques by adding, in all large nations, substantial shares of wind-generated and central solar power, and also (where possible) more geothermal electricity. These requirements would necessitate greatly expanded long-distance high-voltage connections, and this would result in additional large land claims in all countries with extensive territories, but because these claims are difficult to calculate without a great deal of specific assumptions, I will leave them aside.

And although electricity's share of final energy use has been steadily rising (in the United States it is now about 40%), fossil fuels dominate in all major economies, where they have become principal sources for transportation, space heating, and an enormous variety of industrial processes. As a result, their substitution presents a greater transition challenge than displacing significant shares of thermally generated electricity by solar and wind-generated power. Setting aside the early emergence of a hydrogen economy, there are two basic paths toward their replacement. The first one is a complete substitution of fossil fuels by a range of phytomass fuels (wood, liquid biofuels, gasification of phytomass) resembling the current sources and able to fit existing uses.

The second, a more desirable and less land-intensive choice, is to produce some phytomass fuels and substitute a large share of their current use by renewably generated electricity used directly to power electric cars, trains, and many industrial processes, and to provide space heating, and indirectly to produce (with an inevitable reduction of efficiency) storable energies in the form of compressed air, hot water, ice, ammonia, and, of course, hydrocarbons made with captured CO_2 and some hydrogen. Consequently, the most realistic approach to delimit the approximate land requirements of new renewable energy sources is to use relatively extreme but still plausible scenarios of future energy systems that might eventually displace all fossil fuels. Small countries would not have the requisite flexibility of choice, but it would seem that large nations should have plenty of space to eventually put in place complex renewable energy systems. How much land such arrangements would claim would depend on the composition of final energy use.

A higher degree of electrification in sunny and windy places would entail smaller land claims than the large-scale substitution of gasoline, kerosene, and diesel by biofuels. I have sketched two options for the eventual total displacement of fossil fuels for the US demand at the 2012 level, which entailed roughly 320 GW of fuel-generated electricity and 1.8 TW of coal, oil, and gas, with 60% of this fuel total going for transportation. I have assumed fairly high power densities, in all cases higher than today's practices, and with wind-powered electricity generation I am counting only the land actually occupied by pads and access roads. The first option—displacing all fossil fuel–based electricity generation by solar and wind electricity, and

Box 8.1
Land required to displace 2012 US fossil fuel demand by new renewables

Electricity generated from fossil fuels: ~320 GW
   50% displaced by PV (10 W/m^2) = 16,000 km^2
   25% displaced by CSP (20 W/m^2) = 4,000 km^2
   25% displaced by wind (50 W/m^2)* = 1,600 km^2

Liquid fuels**: ~1,100 GW
   25% displaced by crop ethanol (0.4 W/m^2) = ~688,000 km^2
   50% displaced by cellulosic ethanol (0.3 W/m^2) = ~1,830,000 km^2
   25% displaced by biodiesel (0.2 W/m^2) = 1,375,000 km^2

Solid and gaseous fuels***: ~700 GW
   50% displaced by wood combustion (1 W/m^2) = 350,000 km^2
   50% displaced by phytomass gasification (0.8 W/m^2) = ~438,000 km^2

Notes: *Counting only pads and roads. **Nonfuel uses are subtracted. ***Coal and gas used for electricity generation were subtracted.

substituting biofuels for all fossil fuels—would claim about 470 Mha, and the entire system would have a low power density of 0.45 W/m^2, above all as a result of the enormous areas required to produce liquid biofuels.

Massive electrification—with half of all fuels, or about 900 GW, replaced by electricity generated by the same mixture of solar and wind conversions—would reduce the total to about 250 Mha, but as this total does not include losses involved in converting some of the generated electricity to storable fuels, the eventual land claim would be significantly larger. In any case, these approximate calculations delimit plausible extremes: an entirely renewable energy system would occupy roughly 25%–50% of the country's territory (250–470 Mha), compared to about 0.5% (5.5 Mha) of land claimed by today's fossil fuel–hydro-nuclear system.

Of course, there are many plausible adjustments of the grand total, but closer examinations reveal their limits. The land claims for PV-based electricity generation could be cut by a third or a half by placing more panels on roofs rather than amassing them in solar farms in deserts or placing them on abandoned industrial properties or on strips of land along highways. An increasing share of renewable electricity generation can be moved offshore: large marine wind farms already exist (EWEA 2013), and various means of ocean power extraction are under development, ranging from classic ocean thermal energy conversion (OTEC) schemes (Faizal and Ahmed 2012) to various wave-power devices and turbines powered by tides or ocean currents and also including utility-scale undersea energy storage (Slocum et al. 2013).

All of these immersed or submersed conversions share two commonalities: they have to operate in a demanding, corrosive environment, and (with the exception of strong offshore winds, which can be harnessed with high capacity factors) the thermal and kinetic energies they are converting have very low power densities. Even OTEC, the oldest idea among ocean energy conversions, has not progressed to reliably operating projects. Wave devices are in the earliest stages of commercial development, and except in Scotland there are no bold plans for their mass deployment. The demand for liquid fuels can be reduced by a third or more as more efficient vehicles get to dominate the market, and the same is true (in the longer term, owing to a slower turnover of housing stock) of household energy use.

But even after cutting the lower estimate of future US renewable energy needs by a third we are left with nearly 170 Mha (about 17% of the country, and that is without additional transmission ROWs that would be needed to link the new renewable generation capacities), that is, with more land than is in annually harvested crops. Large as it is, such a share could be, in extremis and costs aside, accommodated by the world's third largest nation—but that is not the case when analogical land claim calculations are done for smaller countries or for the largest island nations.

Even under the assumption of a very high average power density of 1 W/m^2 (which would require a higher rate of the system's electrification), the UK would need about 240,000 km^2, or virtually its entire territory. Similarly, McKay (2013, 1) concluded that "in a decarbonized world that is renewable powered, the land area required to maintain today's British energy consumption would have to be similar to the area of Britain." Germany has gone further in installing wind and PV-based electricity-generating capacities than any other affluent economy, but setting up a completely renewable system based on the best available conversions would require, even with a high power density of 1 W/m^2 and even with all roofs covered by PV panels, about 350,000 km^2, again essentially the country's entire area.

But it could be a lot worse. Japan could completely decarbonize with nearly 600,000 km^2 of land devoted to electricity generation and phytomass fuels, nearly 60% more than the area of the four main islands, and the land requirements of fully renewable national energy systems would surpass entire territories for numerous high-energy countries ranging from such island states as Singapore, Taiwan, and Trinidad to such industrial powers as South Korea or the Netherlands. Again, assorted measures (from rooftop PVs and offshore wind to more efficient cars and lights) could cut these demands by a third or a half, but even then those fully renewable systems would claim impractically large shares of national territories. Given these realities, it is astonishing that Lovins (2011b, 40) would claim that "land footprint seems an odd criterion for choosing energy systems: the amounts of land at issue are not large" and that "for civilian energy production, it's merely an intriguing artifact." Some artifact!

Power densities matter, and this means that the transition from predominantly fossil fuel–based to purely renewable energy systems cannot take place—even in affluent, populous countries with large territories and with excellent conditions for PV-based and wind electricity generation and for phytomass cultivation—by simply following a variant of one of the replacement options just outlined. A large-scale international trade in renewables would help, as it does in the modern fossil fuel system, where trade accounts for almost 20% of all coal use, two-thirds of crude oil demand, and nearly one-third of the natural gas supply (BP 2014; Cornot-Gandolphe 2013). But trading low-energy-density phytomass fuels produced with low power densities in Amazonia would not be obviously the same as trading high-energy-density crude oil produced with exceptionally high power densities in the Middle East.

Moreover, the biospheric realities mean that a truly massive trade in phytomass fuels that could be harvested on a large scale and with high yields could be sourced from only a few tropical countries, predominantly Brazil. Extensive new phytomass production in all other countries with large territories (Russia, Canada, the United States, Australia) is limited either by a cold climate or by recurrent droughts, while serious land scarcity eliminates cultivated phytomass as a major option for the world's four most populous low-income nations, China, India, Indonesia, and Pakistan. Marginal lands, barren hilly slopes, and abandoned, low-productivity farmland are claimed to have a large potential for producing biofuels, but, because of their aridity and poor soils, their low productivity (well below 1 W/m^2) would come at a high environmental cost (soil erosion, nutrient loss, biodiversity reduction). A heavy reliance on Brazil would further imperil the always precarious state of Amazonian forests: after several years of decline, the region's deforestation rate rose by nearly 30% between August 2012 and July 2013 (INPE 2013).

Decentralized Energy Supply?

Of course, a purely renewable energy supply would be easier to realize if electricity generation were to be massively decentralized—and decentralization of power and distributed generation have been the leading mantras of renewable energy advocates. The setup entails electricity generation by small units that may or may not be connected to the grid but that are always close to the point of final use, a solution that appeals to green sensibility as it conforms to the small-is-beautiful ideal. However, a reality check is in order: how can this prospect be squared with the growth of megacities whose densely crowded, high-rise blocks may average throughout the year more than 500 W/m^2 and reach 1,000 W/m^2 during the hours of peak demand?

Box 8.2
Decentralized PV generation in Tokyo

Tokyo metropolis	2,186.9 km2
Area of 23 special wards	621.3 km2
Metropolitan energy demand in 2010	723.5 PJ = 22.9 GW
Demand in 23 special wards	~600 PJ = 19 GW
Metropolitan power density	~10 W/m2
Power density in 23 wards	~30 W/m2
Average insolation	154 W/m2
PV efficiency	12%
Performance factor	0.85
Average power density of PV generation	~15 W/m2

Data sources: Tokyo Metropolitan Government (2006, 2012).

Since 2007 more than half of the world's population has been living in cities. By 2050 that share will be above 70%, and more than half will live in megacities with populations of more than 10 million, areas with the highest power density of final energy uses. Even if the power densities of energy use in many megacities were to decline gradually in the decades ahead, it would be impossible to supply them with decentralized PV-based electricity. The world's largest megacity, today's metropolitan Tokyo, offers a perfect example of these limits.

Even if PV panels operating with 12% conversion efficiency were to cover the entirety of more than 600 km^2 of 23 densely populated wards, they could supply only half of Tokyo's energy requirements. And the entire metropolis, whose power density averages about 10 W/m^2, could get all of its energy from PV panels only if they were to cover about 70% of its nearly 2,200 km^2, an impossibility unless we resort to science fiction visions of cities under plastic bubbles. The shares of annual demand that could be realistically delivered are only small fractions of the total. As previously noted, all potentially available roofs within Tokyo's most densely inhabited twenty-three specials wards are about 64 km^2 (Stoll, Smith, and Deinert 2013), that is, just 10% of the total area; but the practical availability would be only a fraction of that, and roofs are a smaller share of a much less densely inhabited area outside the 23 wards.

Extensively used PV panels would thus supply less than 10% of the annual demand of the entire metropolis and less than 1% of the need in its core area, where average power densities are an order of magnitude higher than the metropolitan mean (and could do so only if they were tied to adequate storage capacities). Not surprisingly, the *Global Energy Assessment* for 2012 concluded that local renewables "can therefore only supply urban energy in niche markets (e.g., low-density residential housing), but can provide less than 1% only of a megacity's energy needs" (Global Energy Assessment 2012, 1347). And, obviously, large-scale electricity storage would be essential in order to supply urban infrastructures that operate 24/7/365 and that need to cover demand that often peaks at night owing to household air conditioning in hot climates.

A similar, or even larger, mismatch between the power densities of production and final use will apply to many highly energy-intensive industrial processes, above all to the smelting and casting of metals and to chemical syntheses ranging from ammonia to plastics and composite materials. Obviously, the most efficacious way to supply megacities and energy-intensive industries would be by converting energies with an even higher power density in their proximity, either by relying on domestic resources or by importing high-energy-density fuels—while finding large nearby areas capable of supporting large wind or solar capacities might be impractical or impossible. As there is no such renewable option, megacities would have to rely on high-voltage links to distant concentrations of wind and solar generation capacities.

How disruptive that shift will eventually be to traditional centralized utilities remains uncertain. The Edison Electric Institute concluded that despite the risks presented by the rapidly growing penetration of distributed energy, this shift (as long as its degree remains low, as it does in the United States) is

not currently being discussed by the investment community and factored into the valuation calculus reflected in the capital markets. In fact, electric utility valuations and access to capital today are as strong as we have seen in decades, reflecting the relative safety of utilities in this uncertain economic environment. (Kind 2013, 2)

But a study by the Lawrence Berkeley Laboratory showed substantial revenue erosion once the share of rooftop solar generation reaches 10%, although specific effects of higher shares of solar photovoltaics on earning will depend on the type of utility (Satchwell et al. 2014).

Another way to boost renewable generation would be to have a vastly expanded high-voltage link and eventually a global grid. That has been an aspirational goal for decades (Fuller 1981; GENI 2014), but its early emergence (with links crossing the Bering Strait, and connecting North America with Europe via Iceland) is most unlikely, and future large-scale renewable-source electricity transmission will be limited for a long time to regional interconnections (lines from the Algerian Sahara to the EU, or from Arizona to New England, or from Xinjiang to coastal China). Particularly advantageous links will be those that connect distant production and demand areas across several time zones in order to take advantage of different generation and demand peaks (the three-hour difference between CSP plants in the US Southwest and the populous northeastern coast is perhaps the best example).

At least three other components must come together to make future renewable energy systems, dominated by electricity with an environmentally acceptable share of biofuels, possible: an increased efficiency of all final energy uses, large-scale electricity storage to manage the stochasticity of renewable flows, and an affordable means of using electricity to produce liquid fuels. Considerable improvements in the efficiency of energy conversions would reduce the overall power demand and hence narrow the gap between the power densities of renewable energy conversions and the power densities of common energy uses, making solar and wind much more suitable for a decentralized supply outside megacities.

The availability of mass-scale storage of electricity that could deliver on demand combined blocks of 10^7–10^{10} W would require not only very large numbers of distributed small-scale storages (on the order of 10^3–10^4 W, now commercially available and required in Germany and California) but also the introduction of new forms of storage, with the largest units having capacities of 10^7–10^9 W, something that can be done today only with large hydro stations and that the largest pumped hydro storages can do. The third required component of a renewable energy system is the large-scale conversion of surplus wind and solar electricity generated during peak capacity hours to storable energies, preferably to high-energy-density fuels (best of all to synthetic hydrocarbons) or to hydrogen. Even if all road transport became electrified (an unlikely development anytime soon), fuels would still be needed to power ocean shipping and flying, and hydrocarbons would also be needed for many synthetic processes.

This book has been preoccupied with quantifying fundamental physical qualities, but the pace and extent of energy transitions will be strongly determined by costs and real returns, and a closer look shows that the new renewable energy sources are not exceptionally attractive. This may be a surprising statement given the fact that unsubsidized levelized costs of electricity (LCoe) generated by new renewables is becoming increasingly competitive with the dominant conventional conversions (Lazard 2013; USDOE 2013a). But standard calculations of LCoE do not account for different dispatch characteristics, for potential stranded costs of distributed generation and for social costs affecting those unable to afford distributed generation.

Most notably, published levelized costs have ignored the cost of integration measures at the system level, but without such steps, solar or wind cannot reach large shares of the overall supply. That is why Ueckerdt and co-workers (2013) came up with a new measure that quantifies the system LCoE and includes the integration costs of variable renewable energies (VRE).

Their key finding was that

at moderate wind shares (~20%) integration costs can be in the same range as generation costs of wind power and conventional plants. Integration costs further increase with growing wind shares. We conclude that integration costs can become an economic barrier to deploying VRE at high shares. This implies that an economic evaluation of VRE must not neglect integration costs. (Ueckerdt et al. 2013, 1)

If wind's share of electricity generation reached 20%, its systemwide costs could be 50% higher, and if it reached 40% they might be double the traditional LCoE estimates owing largely to standby power and recurrent overproduction costs. Similarly, an analysis by the Berkeley National Laboratory (Mills and Wiser 2012) found that adding the first 100 MW of solar PV-based electricity to a grid can provide capacity credit equal to 40–70 MW, but as soon as PV's share reaches just 10%, these capacity credits shrink to 20–40 MW. These realities have been inexplicably neglected by the proponents of a rapid energy transition. On the other hand, proponents of PV-based solar and wind energy might argue that the environmental benefits of high shares of renewables make them still the better choice, and, undoubtedly, in some already well-interconnected systems integration costs might be relatively modest

As for the true (that is physical, not monetary) returns, Prieto and Hall (2013) called the attention to the low energy return on investment (EROI) for Spain's solar energy, putting it at no more than 2.45 in terms of thermal equivalents, far below 12–14 needed to maintain the modern civilization. Their findings were strengthened by Weissbach and co-workers (2013) who used a strict exergy concept and updated material databases to compare the EROI of wind, PV, hydro, natural gas, coal, and nuclear power plants on a uniform basis, an approach superior to any used in previous studies.

For the renewables they present both an unbuffered EROI and a buffered return that takes into consideration the cost of storage systems. All systems produce more energy than they consume (EROI greater than 1), but two them are below the economic limit of 7: German solar PV has an unbuffered EROI of 3.9 and a buffered return of 1.6, and corn for energy has an EROI of 3.5. European wind generation has an unbuffered EROI of 16, but its buffered rate (3.9) also falls below the economic threshold. In contrast, combined-cycle gas turbine electricity generation has an EROI of 28, a coal-fired plant has an EROI of 30, and pressurized water reactors have an EROI of 75 (fig. 8.3).

All of this does not mean that a new global energy system based predominantly, if not solely, on conversions of renewable energy flows is impossible. But these realities make it clear that achieving it will be more challenging and will take longer than most of its enthusiastic proponents would have us believe. Technical advances, gradual gains, and fundamental innovations will keep making some of its components more affordable and more efficient, but there is no imminent prospect that they could eliminate the mismatch between the low power densities of stochastic renewable-source energy flows and the relatively high power densities of final energy uses in modern urbanized societies.

Energy studies have accomplished a remarkable feat by largely ignoring space as a key organizing determinant of modern systems supplying fuels and electricity. But space matters: resources occur in specific locations and configurations, and their harnessing, conversion, and use proceed with power densities whose values are fundamentally limited by environmental constants and circumscribed by advancing technical capabilities. Modern

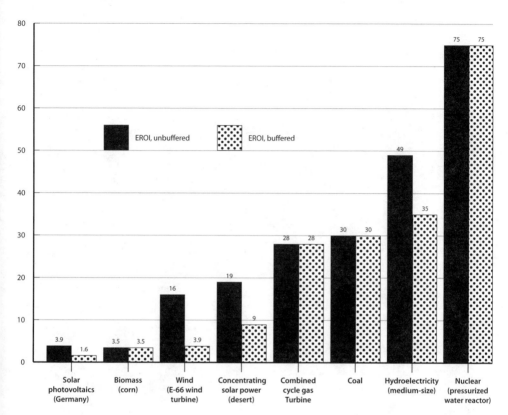

Figure 8.3
EROI of energy production. Modified from Weissbach and co-workers (2013). Carl
De Torres Graphic Design.

civilization has evolved as a direct expression of the high power densities of
fossil fuel extraction, but that extraction is predestined to claim only a
short time span of human evolution. ***New energy arrangements are
both inevitable and desirable, but without any doubt, if they
are to be based on large-scale conversions of renewable energy
sources, then the societies dominated by megacities and concentrated industrial production will require a profound spatial
restructuring of the existing energy system, a process with many
major environmental and socioeconomic consequences.***

Appendix

Basic Système International d'Unités (SI) Units

Quantity	Name	Symbol
Length	meter	m
Mass	kilogram	kg
Time	second	a
Electric current	ampere	A
Temperature	kelvin	K
Amount of substance	mole	mol
Luminous intensity	candela	cd

Energy, Power, and Associated Units

Quantity	Name	Symbol
Electric potential	volt	V
Electric resistance	ohm	Ω
Energy	joule	J
Force	newton	N
Frequency	hertz	Hz
Luminous flux	lumen	lm
Power	watt	W
Electric	W_e	
Installed	W_i	
Peak	W_p	
Thermal	W_t	
Pressure	pascal	Pa
Temperature	degree Celsius	°C

Other Units Used in Calculations and Text

Name and quantity	Symbol
Area swept by turbine blades (m^2)	A
Gravitational constant	g
Hydraulic head, height (in m)	h
Hectare, area (100×100 m)	ha
Mass of water	m
Square meter, area	m^2
Cubic meter, volume	m^3
Energy flux	P
Water flow, volume (in m^3)	Q
Tonne (metric ton), weight (mass)	t
Temperature	T
Wind speed	v
Efficiency (Greek eta, η)	η
Density (Greek rho, ρ)	ρ
Stefan-Boltzmann constant (Greek sigma, σ)	σ

Multiples Used in the Système International d'Unités

Prefix	Abbreviation	Scientific notation
deka	da	10^1
hecto	h	10^2
kilo	k	10^3
mega	M	10^6
giga	G	10^9
tera	T	10^{12}
peta	P	10^{15}
exa	E	10^{18}
zetta	Z	10^{21}
yota	Y	10^{24}

Submultiples Used in the Système International d'Unités

Prefix	Abbreviation	Scientific notation
deci	d	10^{-1}
centi	c	10^{-2}
milli	m	10^{-3}

micro	μ	10^{-6}
nano	n	10^{-9}
pico	p	10^{-12}
femto	f	10^{-15}
atto	a	10^{-18}
zepto	z	10^{-21}
yocto	y	10^{-24}

References

AAR (Association of American Railroads). 2013. Railroads and coal. https://www.aar.org/keyissues/Documents/Background-Papers/Railroads-and-Coal.pdf.

Abandoned Mine Lands Portal. 2013. Abandoned mine lands. http://www.abandonedmines.gov/ep.html.

ABB. 2010. ABB commissions world's longest and most powerful transmission link. http://www.abb.ca/cawp/seitp202/0e95c62a95789c0bc125776500307cd0.aspx.

Abengoa Solar. 2013. Abengoa Solar. http://www.abengoasolar.com/web/en/nuestras_plantas/plantas_en_operacion/espana/PS10_la_primera_torre_comercial_del_mundo.html.

Abu-Zeid, M. A., and F. Z. El-Shibini. 1997. Egypt's High Aswan Dam. *Water Resources Development* 13:209–217.

ACM (American Coalition for Ethanol). 2013. ACE ethanol 101: Frequently asked questions. http://www.ethanol.org/pdf/contentmgmt/EthanolFAQs.pdfA.

Adams, A. S., and D. W. Keith. 2013. Are global wind power resource estimates overstated? *Environmental Research Letters* 8:025021.

Afifi, A. M. 2004. Ghawar: The anatomy of the world's largest oil field. *AAPG Search and Discovery* article no. 20026. http://www.searchanddiscovery.com/documents/2004/afifi01/images/affifi01.pdf.

Alberta Energy. 2013. Natural gas facts. http://www.energy.alberta.ca/NaturalGas/726.asp.

Allan, R. 2011. Warm up to the latest PCB cooling techniques. *Electronic Design,* October 4. http://electronicdesign.com/components/warm-latest-pcb-cooling-techniques.

Allen, L., F. Lindberg, and C.S.B. Grimmond. 2010. Global to city scale urban anthropogenic heat flux: model and variability. *International Journal of Climatology* 13:1990–2005.

Allis, R. G. 1990. Subsidence at Wairakei field. *Transactions (Geothermal Resources Council)* 14:1081–1087.

Alsharhan, A. S., and C.G.D.C. Kendall. 1986. Precambrian to Jurassic rocks of Arabian Gulf and adjacent areas: Their facies, depositional setting, and hydrocarbon habitat. *AAPG Bulletin* 70:977–1002.

Alstom. 2012. Alstom completes Shoaiba III power plant ahead of schedule. http://www.alstom.com/press-centre/2012/5/alstom-completes-shoaiba-iii-power-plant-ahead-of-schedule.

Alyeska Pipeline. 2013. Facts: Trans Alaska Pipeline System. Anchorage, AK: Alyeska Pipeline. http://www.alyeska-pipe.com/TAPS/PipelineFacts.

Amato, I. 2013. Concrete solutions. *Nature* 494:300–301.

American Petroleum Institute. 2009. Hydraulic fracturing operations: Well construction and integrity guidelines. Washington, DC: API. https://www.aar.org/keyissues/Documents/Background-Papers/Railroads-and-Coal.pdf.

Angelo, C. 2012. Growth of ethanol fuel stalls in Brazil. *Nature* 491:646–647.

Arayal, E. 1999. Review of the existing studies related to fuelwood and/or charcoal in Eritrea. Asmara: EC-FAO Partnership Programme. http://www.fao.org/docrep/004/x6788e/x6788e00.HTM.

Arch Coal. 2013. Delivering energy to America. http://www.fao.org/docrep/004/x6788e/x6788e00.HTM.

Archer, C. L., and M. Z. Jacobson. 2005. Evaluation of global wind power. *Journal of Geophysical Research* 110:D12110. doi:10.1029/2004JD005462.

Arnold, B. J., M. S. Klima, and P. J. Bethel, eds. 2007. *Designing the Coal Preparation Plant of the Future*. Englewood, CO: Society for Mining Metallurgy and Exploration.

Arthur, D., and D. Cornue. 2010. Technologies reduce pad size, waste. *American Oil & Gas Reporter,* August 2010. Special report. http://www.all-llc.com/public downloads/AOGR-0810ALLConsulting.pdf.

ATCO Electric. 2010. Facts about direct current transmission lines. http://hvdc.atcoelectric.com/NR/rdonlyres/B4185C91-A96A-46E1-9BCD-30FDA6CBEEF6/0/AtcoFactSheetPROOF6a.pdf.

Athienitis, A. K. 2002. *Thermal Analysis and Design of Passive Solar Buildings*. London: James & James.

Australian Government. 2012. Coal fact sheet. http://www.australianminesatlas.gov.au/education/fact_sheets/coal.html.

AWEO (American Wind Energy Organization). 2013. Areas of industrial wind facilities. http://www.aweo.org/windarea.html.

Azar, K. 2000. The history of power dissipation. *Electronics Cooling* 6 (1): 42–50. http://www.electronics-cooling.com/2000/01/the-history-of-power-dissipation.

Azcárraga, G. 2013. Evaluating the effectiveness of molten salt storage with solar power plants. http://www.ises-online.de/fileadmin/user_upload/PDF/Molten_salt _tower_plant_GA_Azcarraga.pdf.

Bailis, R., C. Rujanevich, P. Dwivedi, et al. 2013. Innovation in charcoal production: A comparative life-cycle assessment of two kiln technologies in Brazil. *Energy for Sustainable Development* 17 (2): 189–200.

Bakker, R. H., E. Pedersen, G. P. van den Berg, R. E. Stewart, W. Lok, and J. Bouma. 2012. Impact of wind turbine sound on annoyance, self-reported sleep disturbance and psychological distress. *Science of the Total Environment* 425:42–51.

Bala, N., Pramod Kumar, N. K. Bohra, N. K. Limba, S. R. Baloch, and G. Singh 2010. Production and decomposition of litter in plantation forests of *Eucalyptus camaldulensis* along canal command area in Indian Desert. *Indian Forester* 36:163–172.

Ball, M., and M. Wietschel, eds. 2009. *The Hydrogen Economy: Opportunities and Challenges*. New York: Cambridge University Press.

Basso, L. C., T. O. Basso, and S. N. Rocha. 2011. Ethanol production in Brazil: The industrial process and its impact on yeast fermentation. In *Biofuel Production: Recent Developments and Prospects,* ed. Marco Aurelio Dos Santos Bernardes. InTech. http://cdn.intechopen.com/pdfs/20058/InTech-Ethanol_production_in_brazil_the _industrial_process_and_its_impact_on_yeast_fermentation.pdf.

Belonger, P., S. E. McKeand, and J. B. Jett. 1997. Wood density assessment of diverse families of loblolly pine using x-ray densitometry. In *Proceedings of the 24th Southern Forest Tree Improvement Conference*, comp. T. White, D. Huber, and G. Powell, 122–142. Orlando, FL, June 9–12, 1997.

Bernardi, M., N. Ferralis, J. H. Wan, et al. 2012. Solar energy generation in three dimensions. *Energy & Environmental Science* 5:6880–6884.

Berndes, G., M. Hoogwijk, and R. van den Broek. 2003. The contribution of biomass in the future global energy supply: A review of 17 studies. *Biomass and Bioenergy* 25:1–28.

Bertani, R. 2010. Geothermal power generation in the world 2005–2010: Update report. In *Proceedings, World Geothermal Congress 2010*. Bali, Indonesia, April 25–29. http://www.energybc.ca/cache/lowtempgeo/Geothermal_electricity_in_the_world _2010_report_Ruggero_Bertani.pdf.

Betz, A. 1926. *Wind-Energie und ihre Ausnutzung durch Windmühlen*. Göttingen: Bandenhoeck & Ruprecht.

Björkman, M. 2004. Long time measurements of noise from wind turbines. *Journal of Sound and Vibration* 277:567–572.

BLM (Bureau of Land Management). 2010. The Geysers geothermal field. http:// www.blm.gov/pgdata/etc/medialib/blm/ca/pdf/pa/energy/geothermal.Par.31107 .File.dat/THE%20GEYSERS%20GEOTHERMAL%20FILED.pdf.

BNSF (Burlington Northern Santa Fe). 2013. Guide to coal mines: Mines served by BNSF Railway. Fort Worth, TX: BNSF Railway. http://www.bnsf.com/customers/pdf/ mineguide.pdf.

Boccard, N. 2009. Capacity factor of wind power: Realized values vs. estimates. *Energy* 37 (7): 2679–2688.

Boisseau, R. 2009. DOE report recommended increasing nuclear waste at Yucca. *AIP Bulletin of Science Policy News* 6 (January 21). http://www.aip.org/fyi/2009/006.html.

Bonano, A., and S. J. Goetz. 2012. Food store density, nutrition education, eating habits and obesity. *International Food and Agribusiness Management Review* 15:1–26.

Bosch Thermotechnology. 2014. Solar thermal flat plate collectors: FKT-1, FKC-1 solar thermal water heating solutions. http://www.bosch-climate.us/files/201205 152058160.Bosch-STS-Collector_FKT_FKC_Sell-Sheet_05.14.12.pdf.

Boxwell, M. 2012. *Solar Electricity Handbook: A Simple Practical Guide to Solar Energy— Designing and Installing Photovoltaic Solar Electric Systems.* Ryton-on-Dunsmore, UK: Greenstream Publishing.

BP (British Petroleum). 2006. Fact sheet Prudhoe Bay. https://dec.alaska.gov/spar/ perp/response/sum_fy06/060302301/factsheets/060302301_factsheet_PB.pdf.

BP. 2014. BP statistical review of world energy. http://www.bp.com/content/dam/ bp/pdf/statistical-review/statistical_review_of_world_energy_2013.pdf.

BPIE (Buildings Performance Institute Europe). 2011. *Europe's Buildings under the Microscope.* Brussels: BPIE. http://www.europeanclimate.org/documents/LR_%20 CbC_study.pdf.

Brandon, N., and D. Thompsett. 2005. *Fuel Cells Compendium.* Amsterdam: Elsevier.

BrightSource. 2013. Ivanpah project facts. http://www.brightsourceenergy.com/ stuff/contentmgr/files/0/8a69e55a233e0b7edfe14b9f77f5eb8d/folder/ivanpah_fact _sheet.10.12.pdf.

Brill, K. G. 2006. 2005–2010 Heat density trends in data processing, computer systems, and telecommunications equipment. http://www.itcrisis.com/pdf/library/ Heat_Density1.pdf.

BSW Solar (Bundesverband Solar Witschaft). 2013. Statistische Zahlen der deutschen Solarwärmebranche (Solarthermie). http://www.solarwirtschaft.de/fileadmin/media/ pdf/2013_2_BSW_Solar_Faktenblatt_Solarwaerme.pdf.

BTS (Bureau of Transportation Statistics). 2013. Oil pipeline profile. http://apps.bts .gov/publications/national_transportation_statistics/html/table_oil_pipeline_profile .html.

Buck, J. L. 1930. *Chinese Farm Economy*. Nanking: University of Nanking.

Buck, J. L. 1937. *Land Utilization in China*. Nanking: University of Nanking.

Bullitt Center. 2013. The greenest commercial building in the world. http:// bullittcenter.org.

Buto, S. G., T. A. Kenney, and S. J. Gerner. 2010. *Land Disturbance Associated with Oil and Gas Development and Effects of Development-Related Land Disturbance on Dissolved-Solids Loads in Streams in the Upper Colorado River Basin, 1991, 2007, and 2025*. Reston, VA: USGS.

Cabadas, J. 2004. *River Rouge: Ford's Industrial Colossus*. Osceola, WI: Motor Books International.

Calpine Corporation. 2013a. Delta Energy Center. http://www.calpine.com/power/ plant.asp?plant=47.

Calpine Corporation. 2013b. Geysers by the numbers. http://www.geysers.com/ numbers.aspx.

Cameco. 2013a. Port Hope. http://www.cameco.com/fuel_and_power/refining_and _conversion/port_hope.

Cameco. 2013b. Uranium operations. http://www.cameco.com.

Cannell, M.G.R. 1989. Physiological basis of wood production: A review. *Scandinavian Journal of Forestry* 4:459–490.

CAPP (Canadian Association of Oil Producers). 2013. Tailing ponds. http://www .oilsandstoday.ca/topics/Tailings/Pages/default.aspx.

Carnegie Steel Company. 1912. Carnegie Steel Company: General statistics and special treatise on Homestead Steel Works. Pittsburgh, PA: Carnegie Steel Company. http://hdl.handle.net/10493/525.

CEA (Central Electricity Authority). 2007. Report on the land requirement of thermal power stations. New Delhi: CEA. http://cea.nic.in/reports/articles/thermal/land _requirement.pdf.

CEGB (Central Electricity Generating Board). 1971. *Modern Power Station Practice: Vol. 1. Planning and Layout*. Oxford: Pergamon Press.

CEPA (Canadian Energy Pipeline Association). 2013. Types of pipelines. http://www .cepa.com/about-pipelines/types-of-pipelines.

Chabot, B. 2013. Renewable electricity in Germany in the first half of 2013. Renewables International, July 8. http://cf01.erneuerbareenergien.schluetersche.de/files/smfiledata/2/8/3/8/2/2/23bREinGH113.pdf.

Chapman, S. 2012. Editorial ignored 17 reviews on wind turbines and health. *British Medical Journal* 344:e3366.

Charlier, R. H., and C. W. Finkl. 2009. *Ocean Energy: Tide and Tidal Power*. Berlin: Springer-Verlag.

Chastain, S. 2009. Pipeline right of way encroachment. *Right of Way*, May/June, 22–27. http://www.irwaonline.org/eweb/upload/0509c.pdf.

Chavalon. 2013. Une centrale électrique au gaz naturel. http://www.chavalon.ch.

Chen, B., G. Shi, B. Wang, J. Zhao, and S. Tan. 2012. Estimation of the anthropogenic heat release distribution in China from 1992 to 2009. *Acta Meteorologica Sinica* 26:507–515.

Cheniere. 2013. Sabine Pass liquefaction project. http://www.cheniere.com/lng_industry/sabine_pass_liquefaction.shtml.

Chincold. 2013. Dam projects. http://www.chincold.org.cn/dams/DamInformation/A2010index_1.htm.

Chubu Electric Power. 2013. Completion of installation of additional LNG tanks at Kawagoe Thermal Power Station. http://www.chuden.co.jp/english/corporate/ecor_releases/erel_pressreleases/3213689_11098.html.

Churkina, G., D. G. Brown, and G. Keoleian. 2010. Carbon stored in human settlements: The coterminous United States. *Global Change Biology* 16:135–143.

Ciais, P., M. Wattenbach, N. Vuichard, et al. 2010. The European carbon balance. Part 2. Croplands. *Global Change Biology* 16:1409–1428.

Clark, D. B., D. A. Clark, and S. Brown. 2002. Stocks and flows of coarse woody debris across a tropical rain forest and topography gradient. *Forest Ecology and Management* 164:237–248.

Clean Energy. 2013a. Agua Caliente Solar Project. http://www.cleanenergyactionproject.com/CleanEnergyActionProject/CS.Agua_Caliente_Solar_Project___Thin_Film_Photovoltaic_Solar_Power_Case_Studies.htm.

Clean Energy. 2013b. Perovo Solar Power Station. http://www.cleanenergyactionproject.com/CleanEnergyActionProject/Solar_PV___Photovoltaic_Solar_Power_Case_Studies_files/Perovo%20Solar%20Power%20Station.pdf.

Clean Energy. 2013c. Sarnia Photovoltaic Power Plant. http://www.cleanenergyactionproject.com/CleanEnergyActionProject/CS.Sarnia_Photovoltaic_Power_Plant___Thin_Film_Photovoltaic_Solar_Power_Case_Studies.html.

Clerici, A. 2007. The Inga development. In *WEC International Forum on Grand Inga Project*. Gaborone, Botswana, March 16–17, 2007. http://89.206.150.89/documents/africainga0307ac_1.pdf.

Cleveland, C., R. U. Ayres, R. Costanza, et al., eds. 2004. *Encyclopedia of Energy*. 6 vols. Amsterdam: Elsevier Academic Press.

Clout, H. 1983. *The Land of France*. London: George Allen & Unwin.

CNI (Confederação Nacional da Indústria). 2012. Forest plantations: Opportunities and challenges for the Brazilian pulp and paper industry on the path of sustainability. Brasília: CNI. https://www.yumpu.com/en/document/view/12277860/forest-plantations-opportunities-and-bracelpa/5.

Cochran, T. B. 2008. How safe is Yucca Mountain? Nashville, TN: Vanderbilt University. http://docs.nrdc.org/nuclear/files/nuc_08010701a.pdf.

Cochran, T. B., H. A. Feiveson, W. Patterson, et al. 20120. Fast breeder reactor programs: History and status. Research Report 8. Princeton, NJ: International Panel on Fissile Materials. http://fissilematerials.org/library/rr08.pdf.

Computer Weekly. 2012. Global census shows datacentre power demand grew 63% in 2012. ComputerWeekly.com, October 8. http://www.computerweekly.com/news/2240164589/Datacentre-power-demand-grew-63-in-2012-Global-datacentre-census.

Copeland, C. 2014. Mountaintop mining: Background on current controversies. Washington, DC: Congressional Research Service. https://www.fas.org/sgp/crs/misc/RS21421.pdf.

Cornot-Gandolphe, S. 2013. Global coal trade: From tightness to oversupply. Paris: Institut Français des Relations Internationales (IFRI). http://www.ifri.org/?page=contribution-detail&id=7570.

Cortez, L. 2011. A preliminary assessment of sugarcane feedstocks in LA and AF (considering the Brazilian experience). Oak Ridge, TN: Oak Ridge National Laboratory. http://web.ornl.gov/sci/ees/cbes/forums/Slides_0512_11.pdf.

Crago, C. L., M. Khanna, J. Barton, et al. 2010. Competitiveness of Brazilian sugarcane ethanol compared to US corn ethanol. Poster presentation, Agriculture & Applied Economics Association Annual Meeting, Denver, CO, July 25–27. http://ageconsearch.umn.edu/bitstream/60895/2/Crago_CostofCornandSugarcaneEthanol_AAEA.pdf.

Cruz, J. 2008. *Ocean Wave Energy: Current Status and Future Perspectives*. Berlin: Springer-Verlag.

Cruz, J. M., and M. S. Taylor. 2012. *Back to the Future of Green Powered Economies*. Cambridge, MA: National Bureau of Economic Research.

CSP World. 2012. Solar one–Solar two. http://www.csp-world.com/cspworldmap/solar-one-solar-two.

CTGC (China Three Gorges Corporation). 2013. Three Gorges Project. http://www.ctgpc.com.

CVSR (California Valley Solar Ranch). 2014. California Valley Solar Ranch. http://www.californiavalleysolarranch.com.

Dabiri, J. O. 2011. Potential order-of-magnitude enhancement of wind farm power density via counter-rotating vertical-axis wind turbine arrays. *Journal of Renewable and Sustainable Energy* 3 (4): 043104.

Data Clean. 2010. Questions and answers on data center cooling issues. http://www.dataclean.com/pdf/CoolingIssuesQA.pdf.

de Beer, J., E. Worrell, and K. Blok. 1998. Future technologies for energy-efficient iron and steel making. *Annual Review of Energy and the Environment* 23:123–205.

de Toma, G., et al. 2004. Solar irradiance variability: Progress in measurement and empirical analysis. *Advances in Space Research* 34:237–242.

de Zeeuw, J. W. 1978. Peat and the Dutch Golden Age: The historical meaning of energy-attainability. *A.A.G. Bijdragen* 21:3–31.

Deane, J. P., B.P.Ó. Gallachóir, and E. J. Mckeogh. 2010. Techno-economic review of existing and new pumped hydro energy storage plant. *Renewable & Sustainable Energy Reviews* 14:1293–1302.

DeBell, D. S., et al. 2002. Wood density and fiber length in young *Populus* stems: Relation to clone, age, growth rate and pruning. *Wood and Fiber Science* 34:529–539.

Degree Days. 2014. Degree heating days. http://www.degreedays.net.

Demographia. 2014. World urban areas. http://www.demographia.com/db-worldua.pdf.

Deng, S., and J. Burnett. 2000. A study of energy performance of hotel buildings in Hong Kong. *Energy and Building* 31:7–12.

Denholm, P., M. Hand, M. Jackson, and S. Ong. 2009. *Land-Use Requirements of Modern Wind Power Plants in the United States*. Golden, CO: National Renewable Energy Laboratory. http://www.nrel.gov/docs/fy09osti/45834.pdf.

Denholm, P., and R. M. Margolis. 2008a. Land-use requirements and the per-capita solar footprint for photovoltaic generation in the United States. *Energy Policy* 36:3531–3543.

Denholm, P., and R. Margolis. 2008b. Supply curves for rooftop solar PV-generated electricity for the United States. Golden, CO: National Renewable Energy Laboratory. http://www.nrel.gov/docs/fy09osti/44073.pdf.

DESERTEC. 2014. The DESERTEC concept. http://www.desertec.org/concept/1144.

Di Matteo, G., et al. 2012. Field performance of poplar for bioenergy in southern Europe after two coppicing rotations: Effects of clone and planting density. *Forest Biogeosciences and Forestry* 5:224–229.

Dickmann, D. I. 2006. Silviculture and biology of short-rotation woody crops in temperate regions: Then and now. *Biomass and Bioenergy* 30:696–705.

Dickson, E. M., J. W. Ryan, and M. H. Smulyan. 1977. *The Hydrogen Economy: A Realistic Appraisal of Prospects and Impacts.* New York: Praeger.

Dickson, M. H., and M. Fanelli. 2004. What is geothermal energy? http://www.geothermal-energy.org/geothermal_energy/what_is_geothermal_energy.html.

Dijkman, T. J., and R. M. J. Benders. 2010. Comparison of renewable fuels based on their land use using energy densities. *Renewable & Sustainable Energy Reviews* 14:3148–3155.

Dinville, A. 2013. Lamberts Point important for trans-Atlantic export. *Worldview* 2 (14): 1.

DiPippo, R. 1991. Geothermal energy: Electricity generation and environmental impact. *Energy Policy* 19:798–807.

Djomo, S. N. 2011. Energy and greenhouse gas balance of bioenergy production from poplar and willow: A review. *Global Change Biology Bioenergy* 3:181–197.

Dodge, E. 2013. Critique of the 100 percent renewable energy for New York plan. http://theenergycollective.com/ed-dodge/301031/critique-100-renewable-energy-new-york-plan.

Domen, F., and A. Jung. 2011. Germany's wind power revolution in the doldrums. *Der Spiegel,* December 30. http://www.spiegel.de/international/business/stress-on-the-high-seas-germany-s-wind-power-revolution-in-the-doldrums-a-805505.html.

Dominion. 2013. Bath County pumped storage station. https://www.dom.com/about/stations/hydro/bath-county-pumped-storage-station.jsp.

Drax Group. 2013. Drax Power Station. http://www.drax.com.

Dubois, M. K. 2010. Hugoton revisited: a new look at an old field. In *New Ways and New Plays: Kansas Oil and Gas Fields.* vol. 5. ed. D. F. Merriam, 50–84. Wichita: Kansas Geological Society.

Duffy, M. 2013. Estimated costs of crop production in Iowa—2014. http://www.extension.iastate.edu/agdm/crops/pdf/a1-20.pdf.

Dukes, J. S. 2003. Burning buried sunshine: Human consumption of ancient solar energy. *Climatic Change* 61:31–44.

Dutta, A., et al. 2011. Process design and economics for conversion of lignocellulosic biomass to ethanol: Thermochemical pathway by indirect gasification and mixed alcohol synthesis. http://www.nrel.gov/biomass/pdfs/51400.pdf.

EC (European Commission). 2006. Photovoltaic solar electricity potential in European countries. http://re.jrc.ec.europa.eu/pvgis/countries/europe/EU-Glob_opta _publications.png.

Elvidge, C., C. Milesi, J. B. Dietz, et al. 2004. U.S. constructed area approaches the size of Ohio. *Eos* 85:233–240.

Elvidge, C. D., B. T. Tuttle, P. C. Sutton, et al. 2007. Global distribution and density of constructed impervious surfaces. *Sensors* 7:1962–1979.

Emerson Network Power. 2009. Energy logic: Reducing data center energy consumption by creating savings that cascade across systems. Columbus, OH: Emerson Network Power. http://www.emersonnetworkpower.com/documents/en-us/latest -thinking/edc/documents/white%20paper/energylogicreducingdatacenterenergy consumption.pdf.

Energi Styrelsen. 2013. Register of wind turbines. http://www.ens.dk/node/2233/ register-wind-turbines.

Enermodal Engineering. 2013. Construction begins on Canada's most energy-efficient office building. http://www.enermodal.com/news/EEL-Newsroom-New-Office.pdf.

Esrafili-Dizaji, B., H. Rahimpour-Bonab, F. K. Harchegani, V. Tavakoli, and M. Naderi. 2013. Great exploration targets in the Persian Gulf: The North Dome/South Pars fields. http://www.findingpetroleum.com/n/Great_exploration_targets_in_the _Persian_Gulf_the_North_DomeSouth_Pars_Fields/ab3518c5.aspx.

Etherington, J. 2009. *The Wind Farm Scam*. London: Stacey International.

Eurostat. 2012. File: Land use, 2010.png. http://epp.eurostat.ec.europa.eu/statistics _explained/index.php?title=File:Land_use,_2010.png&filetimestamp=20121105 110529.

EWEA (European Wind Energy Association). 2013. Deep water: The next step for offshore wind energy. Brussels: EWEA. http://www.ewea.org/fileadmin/files/library/ publications/reports/Deep_Water.pdf.

ExxonMobil. 2008 Baton Rouge refinery. http://corporate.exxonmobil.com/en/ company/worldwide-operations/locations/united-states/baton-rouge.

ExxonMobil. 2013a. Baytown. http://corporate.exxonmobil.com/en/company/ worldwide-operations/locations/united-states/baytown.

ExxonMobil. 2013b. The outlook for energy: A view to 2040. http://corporate .exxonmobil.com/en/energy/energy-outlook.

Faizal, M., and M. R. Ahmed. 2012. On the ocean heat budget and ocean thermal energy conversion. *International Journal of Energy Research* 35:1119.

FAO (UN Food and Agriculture Organization). 2014. FAOSTAT. Rome: FAO. http://faostat.fao.org.

Fargione, J., J. Hill, D. Tilman, S. Polasky, and P. Hawthorne. 2008. Land clearing and the biofuel carbon debt. *Science* 319 (5867): 1235–1238.

Farnham, D. 2001. Corn planting guide. http://www.extension.iastate.edu/publications/pm1885.pdf.

FAS (Federation of American Scientists). 2013. Uranium production. http://www.fas.org/programs/ssp/nukes/fuelcycle/centrifuges/U_production.html.

FCC (Federal Communications Commission). 1996. Guidelines for evaluating the environmental effects of radiofrequency radiation. FCC 96-326. http://transition.fcc.gov/Bureaus/Engineering_Technology/Orders/1996/fcc96326.pdf.

Fiala, J. 2010. Skoda supercritical steam turbine 660 MW. http://e2010.drustvo-termicara.com/resources/files/presentations/fiala.pdf.

Firrisa, M. T. 2011. Energy efficiency for rapeseed biofuel production in different agro-ecological systems. Enschede: University of Twente; http://www.itc.nl/library/papers_2011/msc/gem/firrisa.pdf.

First Hydro Company. 2013. Dinorwig Power Station. http://www.electricmountain.co.uk/en-GB/Dinorwig.

FNR (Fachagentur Nachwachsende Rohstoffe). 2012. *Bioenergy in Germany: Facts and Figures*. Gülzow: FNR.

Foukal, P., et al. 2006. Variations in solar luminosity and their effect on the Earth's climate. *Nature* 443:161–166.

Ford Motor Company. History. n.d. http://www.fordmotorhistory.com/factories/river_rouge/index.php. http://www.fordmotorhistory.com/factories/river_rouge/photos_1.php.

4F Crops. 2011. Future crops for food, feed, fiber and fuel. http://www.4fcrops.eu/pdf/intranet-wp2/D6_Task%202.3&2.4%20CRES.pdf.

Fraunhofer ISE (Fraunhofer Institute for Solar Energy Systems). 2012. Photovoltaics report. Freiburg: Fraunhofer Institute, December 11. http://www.ise.fraunhofer.de/en/downloads-englisch/pdf-files-englisch/photovoltaics-report.pdf.

Fuller, B. 1981. *Critical Path*. New York: St. Martin's Press.

Fthenakis, V. M., and H. C. Kim. 2009. Land use and electricity generation: A life-cycle analysis. *Renewable and Sustainable Energy* 13:1465–1474.

Fthenakis, V. M., and H. C. Kim. 2011. Photovoltaics: Life-cycle analyses. *Solar Energy* 85:1609–1628.

Fthenakis, V. M., and H. C. Kim. 2013. Life cycle assessment of high-concentration photovoltaic systems. *Progress in Photovoltaics: Research and Applications* 21:379–389.

Gagnon, L., C. Bélanger, and Y. Uchiyama. 2002. Life-cycle assessment of electricity generation options: The status of research in year 2001. *Energy Policy* 30:1267–1278.

Galloway, J. A., D. Keene, and M. Murphy. 1996. Fuelling the city: Production and distribution of firewood and fuel in London's region, 1290–1400. *Economic History Review* 49:447–472.

GE Power & Water. 2010. 2.5 MW Wind Turbine Series. http://site.ge-energy.com/prod_serv/products/wind_turbines/en/2xmw/index.htm.

GEA (Geothermal Energy Association). 2013. Geothermal basics: Environment. http://geo-energy.org/geo_basics_environment.aspx.

GENI (Global Energy Network Institute). 2014. GENI.org. http://www.geni.org.

Georgia Power. 2013. Plant Robert W. Scherer. http://www.georgiapower.com/docs/about-us/Plant%20Sherer%20Brochure.pdf.

Geoscience Australia. 2010. Australian energy resource assessment. http://www.ga.gov.au/image_cache/GA16725.pdf.

Gerasimchuk, I., R. Bridle, C. Beaton, and C. Charles. 2012. *State of Play on Biofuel Subsidies: Are Policies Ready to Shift?* Geneva: IISD.

German, J. M. 2004. Hybrid electric vehicles. In *Encyclopedia of Energy,* vol. 3, ed. C. Cleveland, et al., 197–213. Amsterdam: Elsevier Academic Press.

Gerpen, J. V. 2005. Biodiesel processing and production. *Fuel Processing Technology* 86:1097–1107.

GGFR (Global Gas Flaring Reduction Partnership). 2013. Global gas flaring reduction partnership. http://www.flaringreductionforum.org/downloads/GGFR_NewBrochure.pdf.

Ghilardi, A., and F. Steierer. 2011. Charcoal production and use: World country statistics and global trends. Paper presented at "Charcoal Symposium," Arusha, Tanzania, June 15. http://www.charcoalproject.org/wp-content/uploads/2011/08/1_Ghilardi_Steierer_Global_stats.pdf.

GIBSIN Engineers. 2006. Taichung fossil power plant units 1–10. http://www.gibsin.com.tw/old/Case-Taichung1-10FPP.pdf.

Girish, T. E., and S. Aranya. 2012. Photovoltaic power generation on the Moon: Problems and prospects. In *Moon: Prospective Energy and Material Resources,* ed. V. Badescu, 367–376. Berlin: Springer-Verlag.

Global CCS Institute. 2014. Large-scale integrated CCS projects. http://www.globalccsinstitute.com/projects/browse.

Global Energy Assessment. 2012. *Global Energy Assessment.* Cambridge: Cambridge University Press.

Global LNG. 2013. World's LNG liquefaction plants and regasification terminals. http://www.globallnginfo.com/World%20LNG%20Plants%20&%20Terminals.pdf.

Goldman Sachs. 2013. *380 Projects to Change the World: From Resource Constraint to Infrastructure Constraint.* New York: Goldman Sachs.

González-García, S., D. García-Rey, and A. Hospido. 2013. Environmental life cycle assessment for rapeseed-derived biodiesel. *International Journal of Life Cycle Assessment* 18:61–76.

Goodland, R. 1995. The environmentally sustainable challenge for the hydro industry. *Hydropower and Dams* 1:37–42.

Google Earth. 2014. Google Earth. http://www.google.com/earth.

Government of Hong Kong. 2012. Hong Kong energy end-use data 2012. Hong Kong Electrical & Mechanical Services Department. http://www.emsd.gov.hk/emsd/eng/pee/edata.shtml.

Government of Hong Kong. 2013. Land utilization in Hong Kong 2012. Hong Kong Planning Department http://www.pland.gov.hk/pland_en/info_serv/statistic/landu.html.

Grace, J. D., and G. F. Hart. 1990. Urengoy gas field—U.S.S.R. West Siberian Basin, Tyumen District. In *Atlas of Oil and Gas Fields.* Vol. A-109, ed. E. A. Beaumont and N. H. Foster, 309–335. Tulsa, OK: American Association of Petroleum Geologists.

Grand Dixence. 2013. Grand Dixence Dam (altitude 2365 m). http://www.grande-dixence.ch/energie/hydraulic/switzerland/grande-dixence-altitude-2365.html.

Gray, M. R. 2001. Tutorial on upgrading of oils and bitumen. http://www.ualberta.ca/~gray/Links%20&%20Docs/Web%20Upgrading%20Tutorial.pdf.

Greenpeace. 2013. Driving destruction in the Amazon: How steel production is throwing the forest into the furnace. Amsterdam: Greenpeace International. http://www.greenpeace.org/international/Global/international/publications/forests/2012/Amazon/423-Driving-Destruction-in-the-Amazon.pdf.

Grisso, R., M. F. Kocher, and D. H. Vaughan. 2010. Predicting tractor diesel fuel consumption. http://pubs.ext.vt.edu/442/442-073/442-073_pdf.pdf.

Grübler, A. 2004. Transitions in energy use. In *Encyclopedia of Energy,* vol. 6, ed. C. Cleveland, et al., 163–177. Amsterdam: Elsevier Academic Press.

Gunnlaugsson, E. 2012. The Hellisheidi geothermal project: Financial aspects of geothermal development. In *Short Course on Geothermal Development and Geothermal Wells*, UNU-GTP and LaGeo, Santa Tecla, El Salvador, March 11–17, 2012.

Häfele, W., and W. Sassin. 1988. The global energy system. *Annual Review of Energy* 2:1–30.

Hall, K. D., J. Guo, M. Dore, and C. C. Chow. 2009. The progressive increase of food waste in America and its environmental impact. *PLoS ONE* 4 (11): e7940. doi:10.1371/journal.pone.0007940.

Hammersley, G. 1973. The charcoal iron industry and its fuel, 1540–1750. *Economic History Review* 26 (4): 593–613.

Hansen, M. C., P. V. Potapov, R. Moore, et al. 2013. High-resolution global maps of 21st-century forest cover change. *Science* 342:850–853.

Harper, J. A., and J. Kostelnik. 2012. The Marcellus Shale play in Pennsylvania. http://www.marcellus.psu.edu/resources/PDFs/DCNR.pdf.

Harris, J. R. 1988. *The British Iron Industry 1700–1850*. London: Macmillan.

Harvey, G. A. 2008. Simulated reentry heating by torching. NASA Langley Research Center. http://ntrs.nasa.gov/archive/nasa/casi.ntrs.nasa.gov/20080042413_200804 2808.pdf.

Hau, E. 2005. *Wind Turbines: Fundamentals, Technologies, Applications, Economics*. Berlin: Springer-Verlag.

Hausman, W. J. 1980. Business enterprise in 18th-century Britain: Toft Moor Colliery, 1770–79. *Business and Economic History* 9:152–166.

Harris, J. R. 1988. *The British Iron Industry 1700–1850*. London: Macmillan Education.

Hein, F. 2013. Geology of the oil sands. http://www.apega.ca/members/Presentations/ 2013/APEGA%20Geology%20of%20the%20Oil%20Sands.pdf.

Helby, P., H. Rosenqvist, and A. Roos. 2006. Retreat from Salix: Swedish experience with energy crops in the 1990s. *Biomass and Bioenergy* 30:422–427.

Hills Solar. 2008. Brisbane, Queensland collector efficiencies. http://www.energy matters.com.au/renewable-energy/solar-power/solar-hot-water/hills-collector -efficiency.pdf.

Hitachi. 2009. Hitachi develops automotive lithium-ion battery having the world's highest output. Hitachi.com, May 19. http://www.hitachi.com/New/cnews/090519a .html.

Hoffmann, O. 2012. Steel lightweight materials and design for environmental friendly mobility. ThyssenKrupp Steel Europe. http://www.industrialtechnologies 2012.eu/sites/default/files/presentations_session/03_Oliver_Hoffmann.pdf.

Hoogwijk, M., B. de Vries, and W. Turkenburg. 2004. Assessment of the global and regional geographical, technical and economic potential of onshore wind energy. *Energy Economics* 26:889–919.

Hooke, R., B. Le, J. F. Martín-Duque, and J. Pedraza. 2012. Land transformation by humans: A review. *GSA Today* 22:4–10.

Howard, B., J. Parshall, J. Thompson, et al. 2012. Spatial distribution of urban building energy consumption by end use. *Energy and Building* 45:141–151.

Howarth, R. W., and S. Bringezu, eds. 2009. *Biofuels: Environmental Consequences and Interactions with Changing Land Use.* Ithaca, NY: Cornell University.

Hsieh, C., T. Aramaki, and K. Hanaki. 2007. Estimation of heat rejection based on the air conditioner use time and its mitigation from buildings in Taipei City. *Building and Environment* 42:3125–3137.

Hyde, C. K. 1977. *Technological Change and the British Iron Industry 1700–1870.* Princeton, NJ: Princeton University Press.

Hytönen, J. 2008. Intensive production of woody biomass. http://www.tapio.fi/files/tapio/Eng%20sivut/Intesive_production_of_woody_biomass_seminaari.pdf.

IAEA (International Atomic Energy Agency). 2005. Guidebook on environmental impact assessment for in situ leach mining projects. Vienna: IAEA. http://www-pub.iaea.org/MTCD/publications/PDF/te_1428_web.pdf.

IAEA. 2007. Nuclear power plant design characteristics. Vienna: IAEA. http://www-pub.iaea.org/mtcd/publications/pdf/te_1544_web.pdf.

IAEA. 2008. Estimation of global inventories of radioactive waste and other radioactive materials. Vienna: IAEA. http://www-pub.iaea.org/MTCD/publications/PDF/te_1428_web.pdf.

IAEA. 2009. Nuclear fuel cycle information system: A directory of nuclear fuel cycle facilities. Vienna: IAEA. http://www-pub.iaea.org/mtcd/publications/pdf/te_1613_web.pdf.

IAEA. 2012. International status and prospects for nuclear power 2012. Vienna: IAEA. http://www.iaea.org/About/Policy/GC/GC56/GC56InfDocuments/English/gc56inf-6_en.pdf.

IAEA. 2013. *Nuclear Power Reactors in the World.* Vienna: IAEA. http://www-pub.iaea.org/books/IAEABooks/10593/Nuclear-Power-Reactors-in-the-World-2013-Edition.

Iamarino, M., S. Beevers, and C.S.B. Grimmond. 2012. High-resolution (space, time) anthropogenic heat emissions: London 1970–2025. *International Journal of Climatology* 32:1754–1767.

ICAO (International Civil Aviation Organization). 2010. Aviation outlook. Montreal: ICAO. http://www.icao.int/environmental-protection/Documents/Environment Report-2010/ICAO_EnvReport10-Outlook_en.pdf

Ichinose, T., K. Shimodozono, and K. Hanaki. 1999. Impact of anthropogenic heat on urban climate in Tokyo. *Atmospheric Environment* 33:3897–3909.

ICNIRP (International Commission on Non-Ionizing Radiation Protection). 1998. ICNIRP Guidelines for limiting exposure to time-varying electric, magnetic and electromagnetic fields (up to 300 GHz). *Health Physics* 74:494–522.

ICOLD (International Commission on Large Dams). 2014. Register of dams. http://www.icold-cigb.org/GB/World_register/world_register.asp.

IEA (International Energy Agency). 2014. IEA online data services. http://www.iea.org/statistics.

IFRI (International Forestry Resources and Institutions). 2012. IFRI researcher attends a conference about the "4 Fs." http://www.ifriresearch.net/2012/12/20/ifri-researcher-attends-a-conference-about-the-4fs.

IEEE (Institute of Electrical and Electronic Engineers). 2012. National electric safety code. New York: IEEE.

IGA (International Geothermal Association). 2013a. Direct uses. http://www.geothermal-energy.org/geothermal_energy/direct_uses.html.

IGA. 2013b. Installed generating capacity. http://www.geothermal-energy.org/geothermal_energy/electricity_generation.html.

IGU (International Gas Union). 2012. World LNG report 2011. http://www.igu.org/sites/default/files/node-page-field_file/LNG%20Report%202011.pdf.

Ineichen, P. 2011. *Global Irradiance: Average and Typical Year, and Year to Year Annual Variability*. Geneva: Université de Genève. http://www.cuepe.ch/html/biblio/pdf/ineichen_2011_interannual_variability.pdf.

INPE (Instituto Nacional de Pesquisas Espaciais). 2013. Estimativa do Prodes 2013. http://www.obt.inpe.br/prodes/Prodes_Taxa2013.pdf.

Intel. 2005. Intel Pentium 4 processor. http://ark.intel.com/products/27471/Intel-Pentium-4-Processor-551-supporting-HT-Technology-1M-Cache-3_40-GHz-800-MHz-FSB.

Intel. 2013. The story of the Intel 4004. http://www.intel.com/content/www/us/en/history/museum-story-of-intel-4004.html.

IPCC (Intergovernmental Panel on Climate Change). 1995. *IPCC Guidelines for National Greenhouse Gas Inventories*. Paris: IPCC/OECD.

IPCC. 2013. Climate change 2013: The physical science basis. http://www.climatechange2013.org/images/uploads/WGIAR5_WGI-12Doc2b_FinalDraft_Chapter08.pdf.

Ito, A. 2011. A historical meta-analysis of global terrestrial net primary productivity: Are estimates converging? *Global Change Biology.* doi:10.1111/j.1365-2486.2011 .02450.x.

ITTO (International Tropical Timber Organization). 2009. *Encouraging Industrial Forest Plantations in the Tropics.* Yokohama: ITTO.

Jacobson, M. Z. 2008. Review of solutions to global warming, air pollution, and energy security. *Energy and Environmental Science* 2:148–173.

Jacobson, M. Z. 2013. Examining the feasibility of converting New York state's all-purpose energy infrastructure to one using wind, water, and sunlight. *Energy Policy* 57:585–601.

Jacobson, M., and C. L. Archer. 2010. Comment on estimating maximum global land surface wind power extractability and associated climatic consequences. *Earth System Dynamics Discussions* 1:C84–C85.

Jacobson, M. Z., and M. A. Delucchi. 2011. Providing all global energy with wind, water, and solar power. Part I. Technologies, energy resources, quantities and areas of infrastructure, and materials. *Energy Policy* 39:1154–1169.

Jatro. 2014. BioJet fuel. http://biofuel.org.uk/Jatro-BioJet-Fuel.html.

Jeffery, R. D., C. Krogh, and B. Horner. 2013. Adverse health effects of industrial wind turbines. *Canadian Family Physician Medecin de Famille Canadien* 59:473–475.

Johnson, R. C., et al. 2010. An assessment of in-place oil shale resources in the Green River Formation, Piceance Basin, Colorado. Reston, VA: USGS. http://pubs.usgs.gov/ dds/dds-069/dds-069-y/REPORTS/69_Y_CH_1.pdf.

Jordaan, S. M., D. W. Keith, and B. Stelfox. 2009. Quantifying land use of oil sands production: A life cycle perspective. *Environmental Research Letters* 4.024004. http:// iopscience.iop.org/1748-9326/4/2/024004.

Joshi, Y. 2001. Heat out of small packages. *Mechanical Engineering* 123 (12): 56–58.

Juwi Solar. 2008. Neue Dimension der solaren Stromerzeugug. http://www.rio-energie .de/fileadmin/user_upload/Solarenergie/Flyer_Energiepark_Brandis_2008_06.pdf.

Kadam, K. L., and J. D. McMillan. 2003. Availability of corn stover as a sustainable feedstock for bioethanol production. *Bioresource Technology* 88:17–25.

Kapitsa, P. 1976. Physics and the energy problem. *New Scientist* 72 (1021): 10–12.

Kaplan, S. 2007. *Rail Transportation of Coal to Power Plants: Reliability Issues.* Washington, DC: Congressional Research Service.

Kato, M., D. M. DeMarini, A. M. Carvalho, et al. 2005. World at work: Charcoal producing industries in Northeastern Brazil. *Occupational and Environmental Medicine* 62:128–132.

Keith, D. 2013. *A Case for Climate Engineering*. Cambridge, MA: MIT Press.

Khaira, G. 2009. Coal transportation logistics. http://www.tumblerridge.ca/Portals/0/Temp/CN-Oct09.pdf.

Khan, A. A., W. de Jong, P. J. Jansens, and H. Spliethoff. 2009. Biomass combustion in fluidized bed boilers: Potential problems and remedies. *Fuel Processing Technology* 90:21–50.

Kim, E., and S. Barles. 2012. The energy consumption of Paris and its supply areas from the eighteenth century to the present. *Regional Environmental Change* 12:295–310.

Kind, P. 2013. Disruptive challenges: Financial implications and strategic responses to a changing retail electric business. Washington, DC: Edison Electric Institute. http://www.eei.org/ourissues/finance/Documents/disruptivechallenges.pdf.

King, P. 2005. The production and consumption of bar iron in early modern England and Wales. *Economic History Review* 58:1–33.

King, H. 2013. Production and royalties decline in a natural gas well over time. http://geology.com/royalty/production-decline.shtml.

Klasnja, B., S. Orlovic, M. Drekic, and M. Markovic. 2003. Energy production from short rotation poplar plantations. In *VIIth International Symposium Interdisciplinary Regional Research*, 161–166. Hunedoara: ISIRR.

Klein Goldewijk, K. 2001. Estimating global land use change over the past 300 years: The HYDE database. *Global Biogeochemical Cycles* 15:417–433.

Klein Goldewijk, K., A. Beusen, and P. Janssen. 2010. Long term dynamic modeling of global population and built-up area in a spatially explicit way, HYDE 3.1. *Holocene* 20 (4): 565–573.

Kongsager, R., and A. Reenberg. 2012. *Contemporary Land-Use Transitions: The Global Oil Palm Expansion*. Copenhagen: Global Land Project.

Koomey, J. G. 2008. Worldwide electricity used in data centers. *Environmental Research Letters* 3:1–8.

Koomey, J. G. 2011. Growth in data centers electricity use 2005 to 2010. http://www.analyticspress.com/datacenters.html.

Koorey, K. J., and A. D. Fernando. 2010. Concurrent land use in geothermal steam-field development. In *Proceedings, World Geothermal Congress 2010*, 1–6. Bali, April 25–29. http://www.geothermal-energy.org/pdf/IGAstandard/WGC/2010/0207.pdf.

Krausmann, F., and H. Haberl. 2002. The process of industrialization from an energetic metabolism point of view: Socio-economic energy flows in Austria 1830–1995. *Ecological Economics* 41:177–201.

Kumar, A., W. Ma, and G. Gou. 2006. Three Gorges—Shanghai HVDC: Reinforcing interconnection between Central and East China. http://www05.abb.com/global/scot/scot221.nsf/veritydisplay/1a115447a1040f63c125721e0042bdf3/$file/three%20gorges%20-%20shanghai%20hvdc%20-%20%20reinforcing%20interconnection.pdf.

Laderchi, C. R., A. Olivier, and C. Trimble. 2013. *Balancing Act: Cutting Energy Subsidies While Protecting Affordability.* Washington, DC: World Bank.

Lazard. 2013. Lazard's levelized cost of energy analysis: Version 7.0. http://gallery.mailchimp.com/ce17780900c3d223633ecfa59/files/Lazard_Levelized_Cost_of_Energy_v7.0.1.pdf.

Lee, D., V. N. Owens, A. Boe, et al. 2007. *Composition of Herbaceous Biomass Feedstocks.* Brookings, SD: North Central Sun Grant Center.

Lei, Q. 2011. The development of China's cement industry. http://www.tcma.org.tr/images/file/Cin%20Cimento%20Birligi%20Baskani%20Lei%20QIANZHI%20.pdf.

Lenzen, M., and C. J. Dey. 2000. Truncation error in embodied energy analysis of basic iron and steel products. *Energy* 25:577–585.

Lenzen, M., and G. Treloar. 2003. Differential convergence of life-cycle inventories toward upstream production layers, implications for life-cycle assessment. *Journal of Industrial Ecology* 6:3–4.

Leonard, J., III, and J. W. Leonard. 1991. *Coal Preparation.* Littleton, CO: Society for Mining Metallurgy and Exploration.

Li, G. 2011. *World Atlas of Oil and Gas Basins.* Oxford: Wiley-Blackwell.

Likvern, R. 2013. Is the typical NDIC Bakken tight oil well a sales pitch? *The Oil Drum,* April 28, 2013. http://www.theoildrum.com/node/9954.

Lin, X.-Q., D.-F. Zhu, H.-Z. Chen, et al. 2009. Effects of plant density and nitrogen application rate on grain yield and nitrogen uptake of super hybrid rice. *Rice Science* 16:138–142.

Linde. 2010. LNG technology. http://www.linde-engineering.com/internet.global.lindeengineering.global/en/images/LNG_1_1_e_13_150dpi19_4577.pdf.

Liu, W. Y., J. E. D. Fox, and Z. Xu. 2003. Litterfall and nutrient dynamics in a montane moist evergreen broad-leaved forest in Ailao Mountains, SW China. *Plant Ecology* 164:157–170.

Lödl, M., G. Kerber, R. Witzmann, et al. 2010. Abschätzung des Photovoltaik-Potentials auf Dachflächen in Deutschland. In *Symposium Energieinnovation,* Graz, February 10–12, 2010. http://mediatum.ub.tum.de/doc/969497/969497.pdf.

Lorenz, E. N. 1976. *The Nature and Theory of the General Circulation of the Atmosphere.* Geneva: World Meteorological Association.

Lovins, A. B. 1977. *Soft Energy Paths: Toward a Durable Peace*. Cambridge, MA: Friends of the Earth International.

Lovins, A. B. 2011a. *The Essential Amory Lovins*. Ed. C. M. Burns. London: Earthscan.

Lovins, A. B. 2011b. Renewable energy's "footprint" myth. *Electricity Journal* 24 (6): 40–47.

Lu, X., M. B. McElroy, and J. Kiviluoma. 2009. Global potential for wind-generated electricity. *Proceedings of the National Academy of Sciences of the United States of America* 106:10933–10938.

Lund, J., B. Sanner, L. Rybach, et al. 2003. Ground-source heat pumps. *Renewable Energy World* 6:218–227.

Luyssaert, S., P. Ciais, S. L. Piao, et al. 2009. The European carbon balance: Part 3. Forests. *Global Change Biology*. doi:10.1111/j.1365-2486.2009.02056.x.

Madaeni, S. H., R. Sioshansi, and P. Denholm. 2013. Estimating the capacity value of concentrating solar power plants with thermal energy storage: A case study of the Southwestern United States. *IEEE Transactions on Power Systems* 28:1205–1215.

Malone, P. M. 2009. *Waterpower in Lowell: Engineering and Industry in Nineteenth-Century America*. Baltimore, MD: Johns Hopkins University Press.

Manitoba Hydro. 2013a. Limestone generating station. http://www.hydro.mb.ca/corporate/facilities/gs_limestone.shtml.

Manitoba Hydro. 2013b. Transmission and distribution. http://www.hydro.mb.ca/corporate/facilities/transmission_system.shtml.

Marland, G., and M. Obersteiner. 2008. Large-scale biomass for energy, with considerations and cautions: an editorial comment. *Climatic Change* 87:335–342.

Maugeri, L. 2013. *The Shale Oil Boom: A U.S. Phenomenon*. Cambridge, MA: Belfer Center for Science and International Affairs; http://belfercenter.ksg.harvard.edu/files/draft-2.pdf.

Mauthner, F., and W. Weiss. 2013. Solar heat worldwide: Markets and contribution to the energy supply 2011. Gleisdorf, Austria: AEE INTEC. http://www.slideshare.net/UweTrenkner/worldwide2011-ed2013-lo-res.

McDonald, R., J. Fargione, J. Kiesecker, et al. 2009. Energy sprawl or energy efficiency: Climate policy impacts on natural habitat for the United States of America. *PLoS ONE* 4 (8). http://www.plosone.org/article/info:doi/10.1371/journal.pone.0006802.

McKay, A. D., and Y. Miezitis. 2001. Australia's uranium resources, geology and development of deposits. Canberra: AGSO-Geoscience Australia. http://www.ga.gov.au/image_cache/GA9508.pdf.

McKay, D.J.C. 2008. Sustainable energy—Without the hot air. http://www.inference.phy.cam.ac.uk/sustainable/book/tex/cft.pdf.

McKay, D.J.C. 2013. Solar energy in the context of energy use, energy transportation and energy storage. *Philosophical Transactions of the Royal Society A* 371:20110431.

McMahon, M. 2011. Blots on the landscape. *Spectator* 27:18.

McQuaid, J. 2009. Mining the mountains. *Smithsonian*, January. http://www.smithsonianmag.com/ecocenter-energy/mining-the-mountains-130454620.

Mead, D. J. 2005. Forests for energy and the role of planted trees. *Critical Reviews in Plant Sciences* 24:407–421.

Mehani, I., and N. Settou. 2012. Passive cooling of building by using solar chimney. *World Academy of Science, Engineering and Technology* 69:521–525.

Menberg, K. et al. 2013. Long-term evolution of anthropogenic heat fluxes into a subsurface urban heat island. *Environmental Science & Technology* 47:9247–9755.

Methanol Institute. 2013. Methanol facts. http://www.methanol.org/Methanol-Basics/Resources/Methanol-Fuel-Blending-Q-A.aspx.

Michel, J. H. 2005. Status and impacts of the German lignite industry. Göteborg: Swedish NGO Secretariat on Acid Rain. http://www.heuersdorf.de/apc18.pdf.

Mielke, J. E. 1977. Environmental trade-offs of energy supply options. In *Project Interdependence*, 488–555. Washington, DC: Congressional Research Service.

Miller, L. M., F. Gans, and A. Kleidon. 2011. Estimating maximum global land surface wind power extractability and associated climatic consequences. *Earth System Dynamics* 2:1–12.

Miller, R. 2012. Data center racks getting taller, wider, deeper. *Data Center Knowledge*, May 9. http://www.datacenterknowledge.com/archives/2012/05/09/data-center-racks-getting-taller-wider-deeper.

Mills, A., and R. Wiser. 2012. An evaluation of solar valuation methods used in utility planning and procurement processes. http://emp.lbl.gov/sites/all/files/lbnl-5933e_0.pdf.

Minas, L., and B. Ellison. 2009. The problem of power consumption in servers. http://www.drdobbs.com/the-problem-of-power-consumption-in-serv/215800830.

Minerals Council of Australia. 2011. Brown coal: Lignite. http://www.minerals.org.au/file_upload/files/resources/victoria/minerals_fact_sheets/Minerals_-_Fact_Sheets_-_Brown_Coal_-_Lignite.pdf.

Ming, Z., K. Zhang, and D. Liu. 2013. Overall review of pumped-hydro energy storage in China: Status quo, operation mechanism and policy barriers. *Renewable & Sustainable Energy Reviews* 17:35–43.

Mir-Babayev, M. Y. 2002. Azerbaijan's oil history: A chronology leading up to the Soviet era. *Azerbaijan International* 10 (2): 34–40.

Monteiro, M. A. 2006. Em busca do carvão vegetal barato: o deslocamento de siderúrgicas para a Amazônia. *Novos cadernos NAEA* 9 (2): 55–97.

Moreira, J. R. 2006. Global biomass energy potential. *Mitigation and Adaptation Strategies for Global Change* 11:313–342.

Mullan, D. J. 2010. *Physics of the Sun: A First Course.* Boca Raton, FL: CRC Press.

Murthy, V. R., W. van Westrenen, W. Y. Fei. 2003. Experimental evidence that potassium is a substantial radioactive heat source in planetary cores. *Nature* 423:163–165.

NAM (Nederlandse Aardolie Maatschappij). 2009. *Groningen Gas Field Slochteren.* http://www-static.shell.com/content/dam/shell/static/nam-en/downloads/pdf/flyer -namg50eng.pdf.

NASA (National Aeronautic and Space Administration). 2013. Surface meteorology and solar energy. https://eosweb.larc.nasa.gov/sse.

Navigant Consulting. 2009. Ground-source heat pumps: Overview of market status, barriers to adoption, and options for overcoming barriers. http://www1.eere.energy .gov/geothermal/pdfs/gshp_overview.pdf.

NCIB (National Confederation of Industry Brazil). 2012. *Steel Industry in Brazil.* Brasília: NCIB.

Negm, A. M., T. Abdulaziz, M. Nassar and I. Fathy. 2010. Prediction of life time span of High Aswan Dam reservoir using CCHED2D simulation model. In *Fourteenth International Water Technology Conference*, 611–626. Cairo.

Nehring, R. 1978. *Giant Oil Fields and World Oil Resources.* Santa Monica, CA: RAND Corporation.

New York State Department of Environmental Conservation. 2009. SGEIS on the oil, gas and solution mining regulatory program. http://www.dec.ny.gov/energy/47554 .html.

Nilsen, Ø. 2012. Snøhvit introduction to Melkøya plant. http://www02.abb.com/ global/abbzh/abbzh250.nsf/0/56a5a9fc590db243c1257a2100342427/$file/Press_trip _Hammerfest_Presentation_Introduction+to+Melk%C3%B8ya+plant+-+%C 3%98ivind+Nilsen.pdf.

Nissenbaum, M. A., J. J. Aramini, and C. D. Hanning. 2012. Effects of industrial wind turbine noise on sleep and health. *Noise & Health* 14:237–243.

Novinky. 2013. V Česku je extrémně málo slunce, minulý týden svítilo jen dvě hodiny. http://www.novinky.cz/domaci/293799-v-cesku-je-extremne-malo-slunce -minuly-tyden-svitilo-jen-dve-hodiny.html.

Novozymes. 2013. New enzyme technology saves up to 5% corn in ethanol. http:// www.novozymes.com/en/news/news-archive/Pages/New-enzyme-technology-saves -corn-in-ethanol.aspx.

NPS (National Park Service). 2013. Elwha River restoration. http://www.nps.gov/olym/naturescience/elwha-ecosystem-restoration.htm.

NREL (National Renewable Energy Laboratory). 2003. Myths about solar electricity. http://www1.eere.energy.gov/solar/pdfs/32529.pdf.

NREL. 2013. Wind farm area calculator. http://www.energybc.ca/cache/wind2/www.nrel.gov/analysis/power_databook/calc_wind.html.

NREL. 2014. Best research-cell efficiencies. http://www.nrel.gov/ncpv/images/efficiency_chart.jpg.

NYSDEC (New York State Department of Environmental Conservation). 2011. Revised draft SGEIS on the oil, gas and solution mining regulatory program (September). http://www.dec.ny.gov/energy/75370.html.

Odiwe, A. I., and J. I. Muoghalu. 2003. Litterfall dynamics and forest floor litter as influenced by fire in a secondary lowland rain forest in Nigeria. *Tropical Ecology* 44:241–249.

OECD. 2013. *Inventory of Estimated Budgetary Support and Tax Expenditures for Fossil Fuels 2013*. Paris: OECD.

Oil & Gas Journal. 2012. Global oil production up in 2012 as reserves estimates rise again. *Oil & Gas Journal* (December 3):28–31.

Omar, N., M. Daowd, P. van den Bossche, et al. 2012. Rechargeable energy storage systems for plug-in hybrid electric vehicles: Assessment of electrical characteristics. *Energies* 5 (8): 2952–2988.

Osborne, D. 2013. *The Coal handbook: Towards Cleaner Production. Volume 1: Coal production*. Cambridge: Woodhead Publishing.

OST (Office of Science and Technology). 1969. *Considerations Affecting Steam Power Plant Site Selection*. Washington, DC: OST.

Otellini, P. S. 2012. Intel investor meeting 2012. http://www.cnx-software.com/pdf/Intel_2012/2012_Intel_Investor_Meeting_Otellini.pdf.

Pant, M. 2011. Microprocessor power impacts. http://www.lems.brown.edu/~iris/dass11/Pant-DASS.pdf.

Paris, P., L. Mareschi, M. Sabatti, et al. 2011. Comparing hybrid Populus clones for SFR across northern Italy after two biennial rotations: Survival, growth and yield. *Biomass and Bioenergy* 35:1524–1532.

Patterson, R. 2013. Bakken update, July production numbers. *Peak Oil Barrel,* September 13. http://peakoilbarrel.com/eias-international-energy-statistics.

Patzek, T. W. 2006. A statistical analysis of the theoretical yield of ethanol from corn starch. *Natural Resources Research* 15:205–212.

Pauli, E. 2013. Electric power supply in Switzerland. *Current Concerns*, March 5. http://www.currentconcerns.ch/index.php?id=2290.

Paur, J. 2013. In pictures: Solar-powered plane completes epic transcontinental flight. *Wired*, July 10. http://www.wired.com/autopia/2013/07/solar-impulse -gallery.

Peixoto, J. P., and A. H. Oort. 1992. *Physics of Climate*. New York: American Institute of Physics.

Peláez-Samaniego, M. R., M. Garcia-Perez, L. B. Cortez, et al. 2008. Improvements of Brazilian carbonization industry as part of the creation of a global biomass economy. *Renewable & Sustainable Energy Reviews* 12:1063–1086.

Pelamis. 2013. Pelamis technology. http://www.pelamiswave.com/pelamis -technology.

Peng, S., S. Piao, P. Ciais, et al. 2011. Surface urban heat island across 419 global big cities. *Environmental Science & Technology* 46:696–703.

Peplow, M. 2014. Cellulosic ethanol fights for life. *Science* 507:152–153.

Pereira, B.L.C., A. C. Oliveira, A. M. Macedo, et al. 2012. Quality of wood and charcoal from *Eucalyptus* clones for ironmaster use. *International Journal of Forestry Research*. doi:10.1155/2012/523025.

Perez, M.J.R., V. Fthenakis, H.-C. Kim, et al. 2012. Façade-integrated photovoltaics: A life cycle and performance assessment case study. *Progress in Photovoltaics: Research and Applications* 20:975–990.

Perks. R. 2009. Appalachian heartbreak. Washington, DC: Natural Resources Defense Council. http://www.nrdc.org/land/appalachian/files/appalachian.pdf.

Peterson, T. C., and T. W. Owen. 2005. Urban heat island assessment: Metadata are important. *Journal of Climate* 18:2637–2646.

Pfeifer, H. C., L. G. Sousa, and T. T. Silva. 2012. Design of the charcoal blast furnace: Differences to the Coke BF. Paper presented at the 6th International Congress of the Science and Technology in Iron Making. Rio de Janeiro, October 14–18. http://www .abmbrasil.com.br/cim/download/Henrique%20C%20Pleeifer%20Design%20of%20 Charcoal%20Blast.pdf.

Pflueger, J. 2010. Understanding data center energy intensity: A Dell Technical White Paper. http://www.ccbnonprofits.com/images/docs/Dell-Understanding-Data -Center-Energy-Intensity.pdf.

PHMSA (Pipeline and Hazardous Materials Safety Administration). 2013. Annual report mileage for natural gas transmission & gathering systems. http://www.phmsa .dot.gov/portal/site/PHMSA/menuitem.ebdc7a8a7e39f2e55cf2031050248a0c/?vgne

xtoid=35d3f5448a359310VgnVCM1000001ecb7898RCRD&vgnextchannel=3430fb6
49a2dc110VgnVCM1000009ed07898RCRD&vgnextfmt=print.

Pierpont, N. 2009. *Wind Turbine Syndrome: A Report on Natural Experiment.* Santa Fe, NM: K-Selected Books.

Piketty, M.-G., M. Wichert, A. Fallot, and L. Aiomola. 2009. Assessing land availability to produce biomass for energy: The case of Brazilian charcoal for steel making. *Biomass and Bioenergy* 33:180–190.

Pipeline101. 2013. How many pipelines are there? http://www.pipeline101.com/overview/energy-pl.html.

Pollack, H. N., S. J. Hurter, and J. R. Johnson. 1993. Heat flow from the Earth's interior: Analysis of the global data set. *Reviews of Geophysics* 31:267–280.

Potere, D., A. Schneider, S. Angel, and D. L. Civco. 2009. Mapping urban areas on a global scale: Which of the eight maps now available is more accurate? *International Journal of Remote Sensing* 24:6531–6558.

Poynting, J. H. 1884. On the transfer of energy in the electromagnetic Field. *Philosophical Transactions of the Royal Society of London* 175:343–361.

Pretzsch, H. 2009. *Forest Dynamics, Growth and Yield: From Measurement to Model.* Berlin: Springer-Verlag.

Prieto, P. A., and C.A.S. Hall. 2013. *Spain's Photovoltaic Revolution: The Energy Return on Investment.* New York: Springer.

PW Powersystems. 2013. 25 MW Mobile-FT8 gas turbine MOBILEPAC package. http://pwps.com/gas-turbines/25-mw-mobile.html.

Pythagoras Solar. 2014. Solar window. http://www.pythagoras-solar.com/wp-content/uploads/2012/04/product-brochure.pdf.

Qatargas. 2013. Qatargas 1. http://www.qatargas.com/English/QGVentures/Pages/Qatargas1.aspx.

Quah, A.K.L., and M. Roth. 2012. Diurnal and weekly variation of anthropogenic heat emissions in a tropical city, Singapore. *Atmospheric Environment* 46:92–103.

Radartutorial. 2012. Argumentation/derivation. http://www.radartutorial.eu/01.basics/rb13.en.html.

Rao, L.G.G., B. Joseph, and B. Sreemannarayana. 2000. Growth and biomass of some important multipurpose tree species on rainfed sandy loam soils. *Indian Forester* 126:772–781.

Rasmussen, B., I. R. Fletcher, J. J. Brocks, and M. R. Kilburn. 2008. Reassessing the first appearance of eukaryotes and cyanobacteria. *Nature* 455:1101–1104.

Read, P. 2008. Biosphere carbon stock management: Addressing the threat of abrupt climate change in the next few decades. An editorial essay. *Climatic Change* 87:305–320.

Regen, S. W. 2012. Residential buffer zones for wind turbines: The evidence November 2012. http://regensw.s3.amazonaws.com/residential_buffer_zones_for _wind_turbrines_c7cb0ed0aa138678.pdf.

REN21. 2013. *Renewables 2013 Global Status Report.* Paris: REN21 Secretariat.

Renaud, V., and M. Mescall. 2011. Uptime Institute annual report: Data center density. http://symposium.uptimeinstitute.com/schedule/1639-uptime-institute-data -center-density-report.

Repórter Brasil. 2011. Deserto Verde: Os impactos do cultivo de eucalipto e pinus no Brasil. http://escravonempensar.org.br/upfilesfolder/materiais/arquivos/cartilha _deserto%20verde.pdf.

Rio Tinto. 2013. Blair Athol mine. http://www.riotintocoalaustralia.com.au/ ouroperations/321_blair_athol_mine_448.asp.

RiverNet. 2012. Campagne pour le démantelement du barrage de Poutès. http:// www.rivernet.org/general/dams/decommissioning_fr_poutes/poutes_f.htm.

Robelius, F. 2007. Giant oil fields: The highway to oil. Uppsala: Uppsala Universitet. http://uu.diva-portal.org/smash/get/diva2:169774/FULLTEXT01.

Roberts, B. W., D. H. Shepard, K. Caldeira, et al. 2007. Harnessing high-altitude wind power. *IEEE Transactions on Energy Conversion* 22:136–144.

Roberts, J. D., and M. A. Roberts. 2013. Wind turbines: Is there a human health risk? *Journal of Environmental Health* 75:8–17.

Røland, T. H. 2010. Associated petroleum gas in Russia: Reasons for non-utilization. Lysaker: Fridtjof Nansens Institutt. http://www.fni.no/doc&pdf/FNI-R1310.pdf.

Rosenkranz, C. A., U. Köhler, and K. L. Liska. 2011. Modern battery systems for plug-in hybrid electric vehicles. http://www.lifepo4.info/Battery_study/Batteries/Modern _Battery_Systems_for_Plug_In_Hybrid_Vehicles.pdf.

Rossillo-Calle, F., S. Teelucksingh, D. Thran, and M. Seiffert. 2012. *The Potential Role of Biofuels in Commercial Air Transport: Biojetfuel.* London: Imperial College; http:// www.bioenergytrade.org/downloads/T40-Biojetfuel-Report-Sept2012.pdf.

Rousset, P., C. Figueiredo, M. De Souza and W. F. Quirino. 2011. Pressure effect on the quality of eucalyptus wood charcoal for the steel industry: A statistical analysis approach. *Fuel Processing Technology* 92:1890–1897.

Royal Courts of Justice. 2013. Between: The Queen on the application of RWE Npower Renewables Limited claimant—and—Milton Keynes Borough Council

defendant—and—Ecotricity (Next Generation) Limited. http://www.ref.org.uk/Files/
MK_SPD_Judgment.pdf.

Royal Dutch Shell. 2009. Groningen gas field. http://www-static.shell.com/content/
dam/shell/static/nam-en/downloads/pdf/flyer-namg50eng.pdf.

RWE (Rheinisch-Westfälisches Elektrizitätswerk) Power. 2013. Standorte Braunkohle.
http://www.rwe.com/web/cms/de/59998/rwe-power-ag/standorte/braunkohle/
garzweiler.

RWE. 2014. Westfalen power plant. http://www.rwe.com/web/cms/mediablob/en/
331230/data/12428/4/rwe-power-ag/power-plant-new-build/westfalen-power
-station-units-d-and-e/Westfalen-Power-Plant-New-build-project-units-d-and-e.pdf.

Sakai, T., T. Suehiro, and T. Tani. 2009. Geotechnical performance of the Kashiwazaki-Kariwa Nuclear Power Station caused by the 2007 Niigataken Chuetsu-oki
Earthquake. In *Earthquake: Geotechnical Case Histories for Performance-Based Design*,
ed. T. Kokusho, 1–29. Boca Raton, FL: CRC Press.

Sampaio, R. S. 2005. Large-scale charcoal production to reduce CO_2 emissions and
improve quality in the coal based ironmaking industry. Paper presented at the
Workshop and Business Forum on Sustainable Biomass Production for the World
Market, Campinas, November 30–December 3. http://www.bioenergytrade.org/
downloads/sampaionovdec05.pdf.

San Francisco Water Power Sewer. 2013. 2012 Energy benchmarking report.
San Francisco: San Francisco Water Power Sewer. http://sfwater.org/modules/
showdocument.aspx?documentid=4139.

Sandrea, R. 2012. Evaluating production potential of mature US oil, gas shale plays.
Oil & Gas Journal, March 12. http://www.ogj.com/articles/print/vol-110/issue-12/
exploration-development/evaluating-production-potential-of-mature-us-oil.html.

Sarlls, C. E., and G. Oladosu. 2010. A comparison of empirical and theoretical euca-
lyptus yields in Brazil. http://web.ornl.gov/sci/ees/cbes/Trip%20Reports/P2%20
Oladosu%20and%20Sarlis%202010%20POSTER%20Eucalypts%20Productivity%20
Theory%20vs%20Empirical%20%20Compatibility%20Mode.pdf.

Satchwell, A., A. Mills, A. Barbose, et al. 2014. Financial impacts of net-metered PV
on utilities and ratepayers: A scoping study of two prototypical US utilities. http://
emp.lbl.gov/sites/all/files/LBNL%20PV%20Business%20Models%20Report_no%20
report%20number%20(Sept%2025%20revision).pdf.

Saudi Aramco. 2013. Company refineries. http://www.saudiaramco.com/content/
mobile/en/home/our-operations/refining---chemicals/saudi-aramco---owned-refineries
.html?switchToMobile=1.

Schmer, M. R., K. P. Vogel, R. B. Mitchell, R. K. Perrin. 2008. Net energy of cellulosic
ethanol from switchgrass. *Proceedings of the National Academy of Sciences of the United
States of America* 105:464–469.

Schmid & Company. 2010. Protection of water resources from longwall mining is needed in southwestern Pennsylvania. http://www.schmidco.com/Final%20Report%2026%20July%202010.pdf.

Schneider, A., M. A. Friedl, and D. Potere. 2009. Monitoring urban areas globally using MODIS 500m data: New methods based on "urban ecoregions." *Remote Sensing of Environment* 114:1733–1746.

Schneider, M., A. Froggatt, K. Hosokawa, et al. 2013. The world nuclear industry status report 2013. http://www.worldnuclearreport.org/World-Nuclear-Report-2013.html#electricity_generation.

Schulz, M. 2013. Aufstand in der Rotorsteppe. *Der Spiegel*, no. 27. http://www.gegenwind-husarenhof.de/sonstiges/SP_27_13_Aufstand%20in%20der%20Rotorsteppe_Nachdruck.pdf.

Schurr, S. H., and B. C. Netschert. 1960. *Energy in the American Economy 1850–1975*. Baltimore: Johns Hopkins University Press.

Scott, D. C., and J. A. Luppens. 2013. Assessment of coal geology, resources, and reserve base in the Powder River Basin, Wyoming and Montana. U.S. Geological Survey Fact Sheet 2012–3143. http://pubs.usgs.gov/fs/2012/3143.

Searchinger, T., R. Heimlich, R. A. Houghton, et al. 2008. Use of U.S. croplands for biofuels increases greenhouse gases through emissions from land-use change. *Science* 319:1238–1240.

Secretaria de Estado de Meio Ambiente do Pará. 2008. *Instrução Normativa no. 1/2008 de 10/03/2008. Dispõe sobre normas e procedimentos para disciplinar o uso da Guia Florestal GF-PA para o transporte de produtos e/ou subprodutos de origem florestal do Estado do Pará.*

Semprius. 2013. Semprius' 35.5 percent efficiency sets new record for commercially available solar modules. http://www.semprius.com/news/news-releases/semprius-35.5-percent-efficiency-sets-new-record-for-commercially-available-solar-modules.html.

Senes Consultants. 2009. *Port Hope Conversion Facility Site-wide Risk Assessment: Human Health and Ecological Risk Assessment.* Richmond Hill, ON: Senes Consultants.

Sharp. 2013. Sharp develops concentrator solar cell with world's highest conversion efficiency of 44.4%. http://sharp-world.com/corporate/news/130614.html.

Sieferle, R. P. 2001. *The Subterranean Forest: Energy Systems and the Industrial Revolution.* Cambridge: White Horse Press.

Siemens. 2013a. Siemens steam turbines. http://www.energy.siemens.com/hq/en/fossil-power-generation/steam-turbines/?stc=wwecc120618.

Siemens. 2013b. Siemens study: Europe can save EUR 45 billion in its pursuit of renewables. http://www.siemens.com/press/pool/de/pressemitteilungen/2013/ energy/E201305035e.pdf.

Siemens. 2013c. World's biggest gas turbine arrives at Irsching power plant. http:// www.siemens.com/press/en/presspicture/?press=/en/pp_pg/2007/pg200705038_02 _gasturbine_1449801.htm.

Silicon Solar. 2008. SunMaxx™ evacuated solar collectors. http://www.sunmaxxsolar .com/evacuated-tube-solar-collectors.php.

Sinclair, T. R. 2009. Taking measure of biofuel limits. *American Scientist* 97:400–407.

Singh, B. P. 2013. *Biofuel Crops: Production, Physiology and Genetics*. Wallingford, UK: CABI.

Siwakoti, G. 1994. Arun III: Nepal's controversial hydroelectric project. *Environmental Conservation* 21:173.

SJVG (San Joaquin Valley Geology). 2010. Famous gushers of the world. http://www .sjvgeology.org/history/gushers_world.html.

SJVG. 2012. The San Joaquin Valley oil industry. http://www.sjvgeology.org/oil.

Slesser, M. 1978. *Energy in the Economy*. London: Macmillan.

Slocum, A. H., G. E. Fennel, G. Dundar, et al. 2013. Ocean renewable energy storage (ORES) system: Analysis of an undersea energy storage concept. *Proceedings of the IEEE* 101:906–924.

Smeets, E.M.W., and A.P.C. Faaij. 2007. Bioenergy potentials from forestry in 2050. *Climatic Change* 81:353–390.

Smil, V. 1983. *Biomass Energies*. New York: Plenum Press.

Smil, V. 1984. On energy and land. *American Scientist* 72:15–21.

Smil, V. 1987. *Energy Food Environment*. New York: Oxford University Press.

Smil, V. 1991. *General Energetics*. New York: John Wiley.

Smil, V. 1994. *Energy in World History*. Boulder, CO: Westview Press.

Smil, V. 1999. Crop residues: Agriculture's largest harvest. *Bioscience* 49:299–308.

Smil, V. 2003. *Energy at the Crossroads*. Cambridge, MA: MIT Press.

Smil, V. 2004a. *China's Past China's Future*. London: RoutledgeCurzon.

Smil, V. 2004b. Land requirements of energy systems. In *Encyclopedia of Energy*, vol. 3, ed. C. Cleveland et al., 613–622. Amsterdam: Elsevier Academic Press.

Smil, V. 2005. *Creating the Twentieth Century*. New York: Oxford University Press.

Smil, V. 2006. *Transforming the Twentieth Century*. New York: Oxford University Press.

Smil, V. 2008. *Energy in Nature and Society*. Cambridge, MA: MIT Press.

Smil, V. 2010a. *Energy Myths and Realities*. Washington, DC: American Enterprise Institute.

Smil, V. 2010b. *Energy Transitions*. Santa Barbara, CA: Praeger.

Smil, V. 2010c. *Why America Is Not a New Rome*. Cambridge, MA: MIT Press.

Smil, V. 2011. Global energy: The latest infatuations. *American Scientist* 99:212–219.

Smil, V. 2013a. *Harvesting the Biosphere: What We Have Taken from Nature*. Cambridge, MA: MIT Press.

Smil, V. 2013b. *Making the Modern World: Materials and Dematerialization*. Chichester: John Wiley.

Smil, V. 2013c. *Should We Eat Meat?* Chichester: John Wiley.

Smil, V., and K. Kobayashi. 2012. *Japan's Dietary Transition and Its Impacts*. Cambridge, MA: MIT Press.

Smith, J. C. 2013. Wasson Field. In *Handbook of Texas Online*. Texas State Historical Association, http://www.tshaonline.org/handbook/online/articles/downk.

Solar Praxis. 2012. Solarenergie in Deutschland. http://www.seid-2012.com/fileadmin/downloads/SEID12_GW_ES-Screen.pdf.

Solomon, B. D., J. R. Barnes, and K. E. Halvorsen. 2007. Grain and cellulosic ethanol: History, economics, and energy policy. *Biomass and Bioenergy* 31:416–425.

Sorkhabi, R. 2010. The king of giant fields. *GeoExPro* 7:4. http://www.geoexpro.com/article/The_King_of_Giant_Fields/d311f583.aspx.

Sorkhabi, R. 2012. The Great Burgan field, Kuwait. *GeoExpoPro* 9:1. http://www.geoexpro.com/article/The_Great_Burgan_Field_Kuwait/e3fdc547.aspx.

SourceWatch. 2011. The footprint of coal. http://www.sourcewatch.org/index.php/The_footprint_of_coal.

SourceWatch. 2014. Existing U.S. coal plants. http://www.sourcewatch.org/index.php?title=Existing_U.S._Coal_Plants.

SPE (Society of Petroleum Engineers). 1999. Reservoir engineering: Augmented recovery. http://www.spe.org/jpt/print/archives/1999/12/JPT1999_12_reservoir_engrgJPT Series.pdf.

Spitzley, D. V., and G. A. Keoleian. 2004. *Life Cycle Environmental and Economic Assessment of Willow Biomass Electricity: A Comparison with Other Renewable and Non-Renewable Sources*. Ann Arbor, MI: Center for Sustainable Systems, University of Michigan.

Standard & Poor's Rating Services. 2012. After a decade of wind power, the unexpected is still always expected. *Credit Week,* May 23, 45–47.

Steffen, B. 2012. Prospects for pumped-hydro storage in Germany. *Energy Policy* 45:420–429.

Stewart, I. D. 2011. A systematic review and scientific critique of methodology in modern urban heat island literature. *International Journal of Climatology* 31:200–217.

Stewart, D. A., H. P. Dudel, and L. J. Levitt. 1993. *Solar Radiation in Saudi Arabia.* Redstone Arsenal, AL: United States Army Missile Command.

Stiebel Eltron. 2014. SOL 27 Premium Flat Plate Solar Collector. http://www.stiebel -eltron-usa.com/pdf/brochure_sol27.pdf.

Straker, E. 1969. *Wealden Iron.* New York: Augustus M. Kelley.

StratoSolar. 2014. Power platforms. http://www.stratosolar.com.

Stoll, B. L., T. A. Smith, and M. R. Deinert. 2013. Potential for rooftop photovoltaics in Tokyo to replace nuclear capacity. *Environmental Research Letters* 8. doi:10.1088/1748-9326/8/1/014042.

Su, B. 2012. Hotel design and energy consumption. *World Academy of Science. Engineering and Technology* 72:1655–1660.

Subramanian, M. 2012. An ill wind. *Nature* 486:310–311.

Subramanian, M. 2014. Deadly dinners. *Nature* 509:548–551.

SugarCane.org. 2013. Answers to frequently asked questions. http://sugarcane.org/ media-center/faqs

SunPower. 2013. SunPower announces acquisition of Greenbotics, Inc. http:// newsroom.sunpower.com/2013-11-04-SunPower-Announces-Acquisition-of -Greenbotics-Inc

Svenson, C. 2011. Lead acid batteries in micro hybrid and hybrid electrical vehicle applications. http://exide.dk.loopiadns.com/wp/wp-content/uploads/2011/ 01/Svensson_Automassan_2011_01_21.pdf

Swami, S. N., et al. 2009. Charcoal making in the Brazilian Amazon: Economic aspects of production and carbon conversion efficiencies of kilns. In *Amazonian Dark Earths: Wim Sombroek's Vision,* ed. W. I. Woods et al., 411–422. Berlin: Springer-Verlag.

Tabachnyk, M., B. Ehrler, S. Gélinas et al. 2014. Resonant energy transfer of triplet excitons from pentacene to PbSe nanocrystals. *Nature Materials* doi:10.1038/nmat4093.

Teir, S. 2002. *Modern Boiler Types and Applications.* Helsinki: Helsinki University of Technology; http://www.energy.kth.se/compedu/webcompedu/ManualCopy/Steam

_Boiler_Technology/Modern_boiler_applications/modern_boiler_types_and _applications.pdf.

TEPCO (Tokyo Electric Power Company). 2012. *Annual Report 2012*. http://www .tepco.co.jp/en/corpinfo/ir/tool/annual/pdf/ar2012-e.pdf.

Tester, J. W., B. J. Anderson, A. S. Batchelor, et al. 2006. *The Future of Geothermal Energy: Impact of Enhanced Geothermal Systems (EGS) on the United States in the 21st Century*. Cambridge, MA: Massachusetts Institute of Technology; http://mitei.mit .edu/system/files/geothermal-energy-full.pdf.

Thackeray, M. M. 2004. Batteries, transportation applications. In *Encyclopedia of Energy*, vol. 1, ed. C. Cleveland et al., 127–139. Amsterdam: Elsevier Academic Press.

Thermomax Industries. 2010. Collector efficiency. http://www.solarthermal.com/ applications/efficiency/http://www.hsa.ei.tum.de/Publikationen/2010/2010_Loedl _Kerber_Wi_Graz.pdf.

ThyssenKrupp. 2012. Crew celebrates milestone: Europe's biggest blast furnace produces 70 millionth ton of steel. http://www.thyssenkrupp.com/en/presse/art_detail .html&eid=TKBase_1328095396431_284885812.

Tokyo Metropolitan Government. 2006. Environmental White Paper 2006. Tokyo: TMG.

Tokyo Metropolitan Government. 2012. 2011-Nendo tonai no onshitsu kōka gasu haishutsu-ryō oyobi enerugī shōhi-ryō (sokuhō-chi) [Greenhouse gas emissions and energy consumption for the year 2011 in Tokyo (preliminary)]. http://www.kankyo .metro.tokyo.jp/climate/other/attachement/2011sokuho.pdf.

Truax, B., D. Gagnon, J. Fortier, and F. Lambert. 2012. Yield in 8 year-old hybrid poplar plantations on abandoned farmland along climatic and soil fertility gradients. *Forest Ecology and Management* 267:228–239.

Ueckerdt, F., L. Hirth, G. Luderer, and O. Edenhofer. 2013. *System LCOE: What are the costs of variable renewables?* Potsdam: Institute for Climate Impact Research; http://www.pik-potsdam.de/members/Ueckerdt/system-lcoe-working-paper.

Uhlig, A. 2011. Charcoal production in Brazil: Does it pass the sustainability bar? Paper presented at "Charcoal Symposium." Arusha, Tanzania, June 15. http://www .charcoalproject.org/wp-content/uploads/2011/08/10_Uhlig_Brazil_stats.pdf.

Umov, N. A. 1874. Ein Theorem über die Wechselwirkungen in Endlichen Entfernungen. *Zeitschrift für Mathematik und Physik* 19:97.

UN (United Nations). 2012. *World Urbanization Prospects: The 2011 Revision*. New York: UN.

UN. 2014. Energy statistics. http://unstats.un.org/unsd/energy.

USDA. 2013a. 2012 agricultural statistics annual. http://www.nass.usda.gov/ Publications/Ag_Statistics/2012/index.asp.

USDA (US Department of Agriculture). 2013b. Food availability (per capita) data system. http://www.ers.usda.gov/data-products/food-availability-(per-capita) -data-system.aspx.

USDOE (US Department of Energy). 1983. *Energy Technology Characterizations Handbook: Environmental Pollution and Control Factors.* Washington, DC: USDOE.

USDOE. 1998. Solar Two. http://www.nrel.gov/docs/fy99osti/24643.pdf.

USDOE. 2005. Biomass as feedstock for a bioenergy and bioproducts industry: The technical feasibility of a billion-ton annual supply. http://www1.eere.energy.gov/ bioenergy/pdfs/final_billionton_vision_report2.pdf.

USDOE. 2011a. U.S. billion-ton update: Biomass supply for a bioenergy and bioproducts industry. http://www1.eere.energy.gov/bioenergy/pdfs/billion_ton_update.pdf.

USDOE. 2011b. Wind powering America. http://www.windpoweringamerica.gov/ policy/ordinances.asp.

USDOE. 2013a. Annual energy outlook 2013. http://www.eia.gov/forecasts/aeo/ pdf/0383(2013).pdf.

USDOE. 2013b. Buildings energy data book. http://buildingsdatabook.eren.doe.gov.

USDOE. 2014. Enhanced geothermal systems. http://energy.gov/eere/geothermal/ enhanced-geothermal-systems-0.

USEIA (US Energy Information Agency). 1993. The changing structure of the electric power industry, 1970–1991. http://webapp1.dlib.indiana.edu/virtual_disk_library/ index.cgi/4265704/FID3754/pdf/electric/0562.pdf.

USEIA. 2007. About U.S. natural gas pipelines. http://www.eia.gov/pub/oil_gas/ natural_gas/analysis_publications/ngpipeline/fullversion.pdf.

USEIA. 2009. Residential energy consumption survey (RECS). http://www.eia.gov/ consumption/residential/data/2009/index.cfm?view=microdata.

USEIA. 2011a. Annual coal report. December 12. http://www.eia.gov/coal/annual/ index.cfm.

USEIA. 2011b. Over one-third of natural gas produced in North Dakota is flared or otherwise not marketed. http://www.eia.gov/todayinenergy/detail.cfm?id=4030.

USEIA. 2011c. Review of emerging resources: U.S. shale gas and shale oil plays. http://www.eia.gov/analysis/studies/usshalegas.

USEIA. 2012. Crude oil production and crude oil well productivity, 1954–2011. http://www.eia.gov/totalenergy/data/annual/showtext.cfm?t=ptb0502.

USEIA. 2013a. Annual energy outlook 2013 with projections to 2040. http://www .eia.gov/forecasts/aeo/pdf/0383(2013).pdf.

USEIA. 2013b. Electricity. http://www.eia.gov/electricity.

USEIA. 2013c. Natural gas gross withdrawals and production. http://www.eia.gov/ dnav/ng/ng_prod_sum_a_epg0_fgw_mmcf_a.htm.

USEIA. 2013d. Nuclear and uranium. http://www.eia.gov/nuclear.

USEIA. 2014a. Consumption of kerosene. http://www.eia.gov/cfapps/ipdbproject/ iedindex3.cfm?tid=5&pid=64&aid=2&cid=regions&syid=2008&eyid=2012&unit=T BPD.

USEIA. 2014b. Existing transmission capacity by high-voltage size, 2012. http:// www.eia.gov/electricity/annual/html/epa_08_10_a.html.

USEIA. 2014c. Natural gas. http://www.eia.gov/naturalgas/

USEIA. 2014d. Net generation by energy source. http://www.eia.gov/electricity/ monthly/epm_table_grapher.cfm?t=epmt_1_1

USGS (United States Geological Survey). 2000. National land cover dataset. U.S. Geological Survey fact sheet 108–00. http://erg.usgs.gov/isb/pubs/factsheets/fs10800 .html.

USGS. 2013. Coal resources of selected coal beds and zones in the northern and central Appalachian basin. http://pubs.usgs.gov/fs/fs004-02/fs004-02.pdf.

USNRC (US Nuclear Regulatory Commission). 2012a. Generic environmental impact statement for license renewal of nuclear plants. NUREG-1437. Vol. 1. http://www .nrc.gov/reading-rm/doc-collections/nuregs/staff/sr1437.

USNRC. 2012b. Locations of independent spent fuel storage installations. http:// www.nrc.gov/waste/spent-fuel-storage/locations.html.

USNRC. 2013. Crow Butte resources site. http://www.nrc.gov/info-finder/materials/ uranium/licensed-facilities/crow-butte.html.

Vatenfall. 2013. Jänschwalde. http://powerplants.vattenfall.com/node/288.

Vestas. 2013. V 90 3.0 MW. http://nozebra.ipapercms.dk/Vestas/Communication/ Productbrochure/V9030MW/V9030MWUK.

Voith. 2011. Rheinfelden, Germany/Switzerland. http://www.voith.com/en/markets -industries/industries/hydro-power/modernization/rheinfelden_main-11453-11453 .html.

Volta River Authority. 2013. Akosombo hydro plant. http://vraghana.com/our _mandate/akosombo_hydro_plant.php.

Waffenschmidt, E. 2008. Die Rolle der Photovoltaik bei einer 100-Prozent Versorgung Deutschlands mit erneuerbaren Energien. http://www.waffenschmidt.homepage.t-online.de/documents/100prozent-solar-staffelstein_CD_final.pdf.

Warde, P. 2007. *Energy Consumption in England and Wales, 1560–2004*. Napoli: Consiglio Nazionale della Ricerche.

Weissbach, D., G. Ruprecht, A. Huke, et al. 2013. Energy intensities, EROIs (energy returned on invested), and energy payback times of electricity generating power plants. *Energy* 52:210–221.

West, P. W. 2013. *Growing Plantation Forests*. Berlin: Springer.

Whitelaw, E., and E. MacMullan. 2002. A framework for estimating the costs and benefits of dam removal. *Bioscience* 52:724–730.

Wirth, H. 2013. Recent facts about photovoltaics in Germany. http://www.ise.fraunhofer.de/en/publications/veroeffentlichungen-pdf-dateien-en/studien-und-konzeptpapiere/recent-facts-about-photovoltaics-in-germany.pdf.

Wiser, R., and M. Bolinger. 2012. 2011 wind technologies market report. http://www1.eere.energy.gov/wind/pdfs/2011_wind_technologies_market_report.pdf.

Wiser, R., and M. Bolinger. 2013. 2012 wind technologies market report. http://www1.eere.energy.gov/wind/pdfs/2012_wind_technologies_market_report.pdf.

WNA (World Nuclear Association). 2010. Geology of uranium deposits. http://www.world-nuclear.org/info/Nuclear-Fuel-Cycle/Uranium-Resources/Geology-of-Uranium-Deposits.

WNA. 2012. The nuclear fuel cycle. http://www.world-nuclear.org/info/Nuclear-Fuel-Cycle/Introduction/Nuclear-Fuel-Cycle-Overview.

WNA. 2013a. Australia's uranium mines. http://www.world-nuclear.org/info/Country-Profiles/Countries-A-F/Appendices/Australia-s-Uranium-Mines.

WNA. 2013b. World nuclear power reactors and uranium requirements. http://www.world-nuclear.org/info/Facts-and-Figures/World-Nuclear-Power-Reactors-and-Uranium-Requirements/#.UmmEIvco6M8.

WNA. 2013c. World uranium mining production. http://www.world-nuclear.org/info/Nuclear-Fuel-Cycle/Mining-of-Uranium/World-Uranium-Mining-Production.

Wong, K. V., Y. Dai, and B. Paul. 2012. Anthropogenic heat release into the environment. *Journal of Energy Resources Technology* 134 (4): 041602–041607. doi:10.1115/1.4007360.

Wong, K. V., A. Paddon, and A. Jimenez. 2013. Review of world urban heat islands: Many linked to increased mortality. *Journal of Energy Resources Technology* 135:022101–022111.

World Coal Association. 2013. Coal and electricity. http://www.worldcoal.org/coal/uses-of-coal/coal-electricity.

World Steel Association. 2013. Statistics archive. http://www.worldsteel.org/statistics/statistics-archive.html.

Worley, M., and J. Yale. 2012. Biomass gasification technology assessment. Golden, CO: NREL; http://www.nrel.gov/docs/fy13osti/57085.pdf.

Wyoming State Geological Society. 2013. Coal production and mining. http://www.wsgs.uwyo.edu/research/energy/coal/Production-Mining.aspx.

Yeh, S., S. M. Jordaan, A. R. Brandt, et al. 2010. Land use greenhouse gas emissions from conventional oil production and oil sands. *Environmental Science & Technology* 44:8766–8772.

Zebroski, E., and M. Levenson. 1976. The nuclear fuel cycle. *Annual Review of Energy* 1:101–130.

Zhang, Z., and S. Cotterill. 2008. China's first AC powered walking dragline in coal mining. https://mining.cat.com/cda/files/2793951/7/scotterill%20paper%20minexpo%20final.pdf.

Zhao, M., F. A. Heinsch, R. R. Nemani, et al. 2005. Improvements of the MODIS terrestrial gross and net primary production global data set. *Remote Sensing of Environment* 95:164–176.

Zhou, Y., P. Luckow. S. J. Smith and L. Clarke. 2012. Evaluation of global onshore wind energy potential and generation costs. *Environmental Science & Technology* 46:7857–7864.

Index